Advanced Topics in Welding: Aluminum

Trainee Guide
Fourth Edition

Prentice Hall
Boston Columbus Indianapolis New York San Francisco Upper Saddle River
Amsterdam Cape Town Dubai London Madrid Milan Munich Paris Montreal Toronto
Delhi Mexico City Sao Paulo Sydney Hong Kong Seoul Singapore Taipei Tokyo

National Center for Construction Education and Research

President: Don Whyte
Director of Product Development: Daniele Stacey
Welding Project Manager: Daniele Stacey
Production Manager: Tim Davis
Quality Assurance Coordinator: Debie Ness
Editors: Rob Richardson, Matt Tischler
Desktop Publishing Coordinator: James McKay
Production Assistant: Laura Wright

NCCER would like to acknowledge the contract service provider for this curriculum:
Topaz Publications, Syracuse, New York.

12 13 14 15 16

Prentice Hall
is an imprint of

www.pearsonhighered.com

ISBN 10: 0-13-213722-4
ISBN 13: 978-0-13-213722-5

Preface

To the Trainee

Congratulations! If you're training under an NCCER-Accredited Training Sponsor, you have likely completed *Welding Level Three* and mastered welding of carbon and stainless steel on plate and pipe. Now you are ready to weld the most difficult alloy of all—aluminum.

In *Advanced Topics in Welding: Aluminum*, you'll learn how to set up GMAW and GTAW equipment for welding aluminum plate and pipe. You'll learn how to clean and prepare aluminum coupons for welding and how to avoid problems often encountered in aluminum welds. You'll also learn how to make fillet welds on aluminum plate in 1F, 2F, 3F, and 4F positions and how to make V-groove welds on aluminum plate with backing in the 1G, 2G, 3G, and 4G positions.

On aluminum pipe, you'll learn how to make V-groove welds in the 2G, 5G, and 6G positions.

Learning these advanced aluminum welding techniques will better equip you to compete for these high-demand positions in the industry.

Features of This Book

In addition to the engaging features and sidebars presented in the last edition, Going Green features have been incorporated to highlight energy conservation, as well as practices and products that have lesser environmental impacts.

This edition has been resequenced to contain all aluminum plate and pipe welding processes to include GMAW and GTAW plate welding in the 1F, 2F, 3F, and 4F positions. The GMAW and GTAW pipe welding modules instruct welds in the 1G, 2G, 3G, and 4G.

This edition also correlates to the American Welding Society's (AWS) SENSE EG3.0-96 and EG4.0-96 standards and guidelines. This means that, in addition to conforming to NCCER guidelines for credentialing through its National Registry, this program can also be used to meet guidelines provided by AWS for Advanced and Expert Welder training. For more information on the AWS SENSE program, contact AWS at 800-443-9353 or visit www.aws.org. For information on NCCER's Accreditation and National Registry, contact NCCER Customer Service at 1-888-622-3720 or visit www.nccer.org.

We invite you to visit the NCCER website at www.nccer.org for the latest releases, training information, newsletter, and much more. You can also reference the Contren® product catalog online at www.nccer.org. Your feedback is welcome. You may e-mail your comments to curriculum@nccer.org or send general comments and inquiries to info@nccer.org.

Contren® Learning Series

The National Center for Construction Education and Research (NCCER) is a not-for-profit 501(c)(3) education foundation established in 1995 by the world's largest and most progressive construction companies and national construction associations. It was founded to address the severe workforce shortage facing the industry and to develop a standardized training process and curricula. Today, NCCER is supported by hundreds of leading construction and maintenance companies, manufacturers, and national associations. The Contren® Learning Series was developed by NCCER in partnership with Pearson Education, Inc., the world's largest educational publisher.

Some features of NCCER's Contren® Learning Series are as follows:

- An industry-proven record of success
- Curricula developed by the industry for the industry
- National standardization providing portability of learned job skills and educational credits
- Compliance with Office of Apprenticeship requirements for related classroom training (CFR 29:29)
- Well-illustrated, up-to-date, and practical information

NCCER also maintains a National Registry that provides transcripts, certificates, and wallet cards to individuals who have successfully completed modules of NCCER's Contren® Learning Series. *Training programs must be delivered by an NCCER Accredited Training Sponsor in order to receive these credentials.*

Contren® Curricula

NCCER's training programs comprise nearly 80 construction, maintenance, pipeline, and utility areas and include skills assessments, safety training, and management education.

Boilermaking
Cabinetmaking
Carpentry
Concrete Finishing
Construction Craft Laborer
Construction Technology
Core Curriculum: Introductory Craft Skills
Drywall
Electrical
Electronic Systems Technician
Heating, Ventilating, and Air Conditioning
Heavy Equipment Operations
Highway/Heavy Construction
Hydroblasting
Industrial Coating and Lining Application Specialist
Industrial Maintenance Electrical and Instrumentation Technician
Industrial Maintenance Mechanic
Instrumentation
Insulating
Ironworking
Masonry
Millwright
Mobile Crane Operations
Painting
Painting, Industrial
Pipefitting
Pipelayer
Plumbing
Reinforcing Ironwork
Rigging
Scaffolding
Sheet Metal
Site Layout
Sprinkler Fitting
Tower Crane Operator
Welding

Pipeline

Control Center Operations, Liquid
Corrosion Control
Electrical and Instrumentation
Field Operations, Liquid
Field Operations, Gas
Maintenance
Mechanical

Energy

Introduction to the Power Industry
Power Industry Fundamentals
Power Generation Maintenance Electrician
Power Generation I&C Maintenance Technician
Power Generation Maintenance Mechanic
Steam and Gas Turbine Technician
Introduction to Solar Photovoltaics
Introduction to Wind Energy

Safety

Field Safety
Safety Orientation
Safety Technology

Management

Introductory Skills for the Crew Leader
Project Management
Project Supervision

Supplemental Titles

Applied Construction Math
Careers in Construction
Tool for Success
Your Role in the Green Environment

Spanish Translations

Basic Rigging (Principios Básicos de Maniobras)
Carpentry Fundamentals (Introducción a la Carpintería)
Carpentry Forms (Carpintería de Formas)
Concete Finishing Level One (Acabado de Concreto, Nivel Uno)
Core Curriculum: Introductory Craft Skills (Currículo Básico: Habilidades Introductorias del Oficio)
Drywall Level One (Paneles de Yeso, Nivel Uno)
Electrical Level One (Electricidad, Nivel Uno)
Field Safety (Seguridad de Campo)
Insulating Level One (Aislamiento, Nivel Uno)
Masonry Level One (Albañilería, Nivel Uno)
Pipefitting Level One (Instalación de Tubería Industrial, Nivel Uno)
Reinforcing Ironwork Level One (Herreria de Refuerzo, Nivel Uno)
Safety Orientation (Orientación de Seguridad)
Scaffolding (Andamios)
Sprinkler Fitting Level One (Instalación de Rociadores, Nivel Uno)

Acknowledgments

This curriculum was revised as a result of the farsightedness and leadership of the following sponsors:

Alabama Southern Community College
Applied Welding Technology
B E & K Construction Company
Baran Tech
CNM Community College
Community College of NMSU
Gulf States, Inc. Midland, MI
Lee College
Lincoln Tech
Lousiana Dept. of Education
Northland Pioneer College
Plant Outage Solution, LLC
Spec-Weld Technologies
Sunoco, Inc.
TIC Holdings, Inc.
Tulsa Technology Center
University of Alaska Southeast
Zachry Industrial, Inc.

This curriculum would not exist were it not for the dedication and unselfish energy of those volunteers who served on the Authoring Team. A sincere thanks is extended to the following:

Don Bancrot
Steve Brandow
Barry Breeden
Curtis Casey
Bill D. Cherry
Sheldon Ellis
Drew Fontenot
John Gault
Travis Hamblet
Kay Hamby
Rod Hellyer
Frank D. Johnson
John Knapp
Paul LaBorde
Terry Lowe
Brian McIntosh
Jeff D. Morris
Joseph Murlin
D. Larry Thurston
David L. Twitty

NCCER Partners

American Fire Sprinkler Association
Associated Builders and Contractors, Inc.
Associated General Contractors of America
Association for Career and Technical Education
Association for Skilled and Technical Sciences
Carolinas AGC, Inc.
Carolinas Electrical Contractors Association
Center for the Improvement of Construction Management and Processes
Construction Industry Institute
Construction Users Roundtable
Design Build Institute of America
Merit Contractors Association of Canada
Metal Building Manufacturers Association
NACE International
National Association of Manufacturers
National Association of Minority Contractors
National Association of Women in Construction
National Insulation Association
National Ready Mixed Concrete Association
National Technical Honor Society
National Utility Contractors Association
NAWIC Education Foundation
North American Technician Excellence
Painting & Decorating Contractors of America
Portland Cement Association
SkillsUSA
Steel Erectors Association of America
U.S. Army Corps of Engineers
Women Construction Owners & Executives, USA
University of Florida, M.E. Rinker School of Building Construction

Contents

GMAW – Aluminum Plate

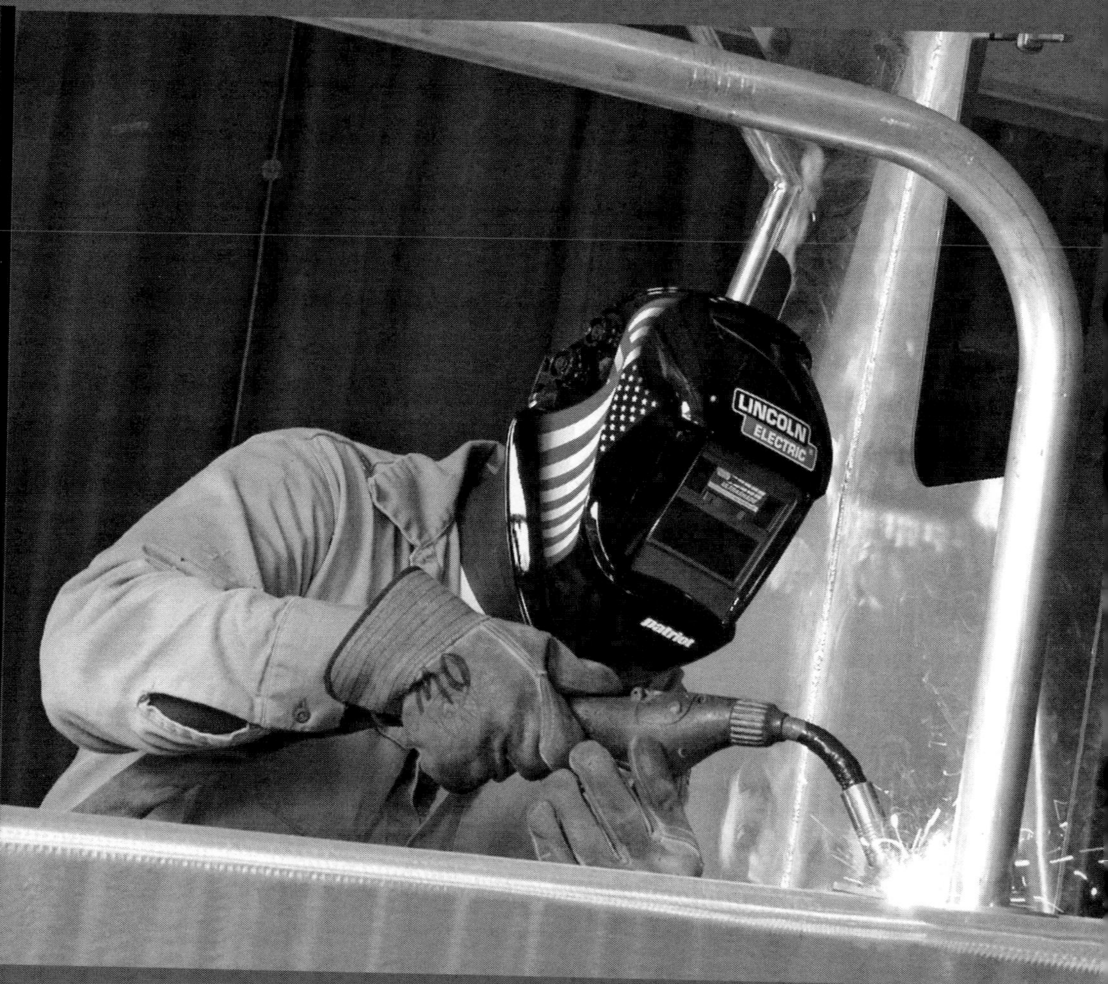

29401-10

Advanced Topics in Welding: Aluminum

29404-10 GMAW – Aluminum Pipe

29403-10 GTAW – Aluminum Pipe

29402-10 GTAW – Aluminum Plate

29401-10 GMAW – Aluminum Plate

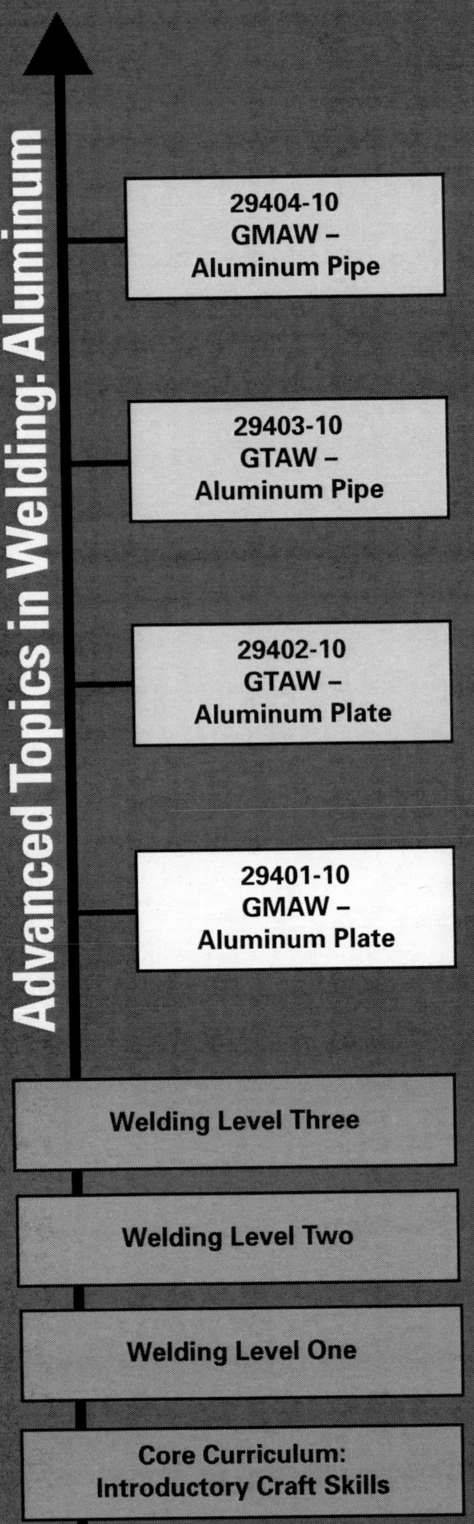

Welding Level Three

Welding Level Two

Welding Level One

Core Curriculum: Introductory Craft Skills

This course map shows all of the modules in *Advanced Topics in Welding: Aluminum.* The suggested training order begins at the bottom and proceeds up. Skill levels increase as you advance on the course map. The local Training Program Sponsor may adjust the training order.

Objectives

When you have completed this module, you will be able to do the following:

1. Identify and explain aluminum metallurgy and the characteristics of aluminum.
2. Explain GMAW and how to set up GMAW equipment to weld aluminum.
3. Build a pad with stringer beads and weave beads, using GMAW equipment, aluminum wire, and shielding gas.
4. Perform multiple-pass fillet welds on aluminum plate in the following positions, using GMAW equipment, aluminum wire, and shielding gas:
   - 1F
   - 2F
   - 3F
   - 4F
5. Perform multiple-pass V-groove welds on aluminum plate with backing in the following positions, using GMAW equipment, aluminum wire, and shielding gas:
   - 1G
   - 2G
   - 3G
   - 4G

Trade Terms

Anodic coating
Coalescence
Ingot
Liquation

Prerequisites

Before you begin this module, it is recommended that you successfully complete *Core Curriculum*; *Welding Level One*; *Welding Level Two*; and *Welding Level Three.*

Contents

Topics to be presented in this module include:

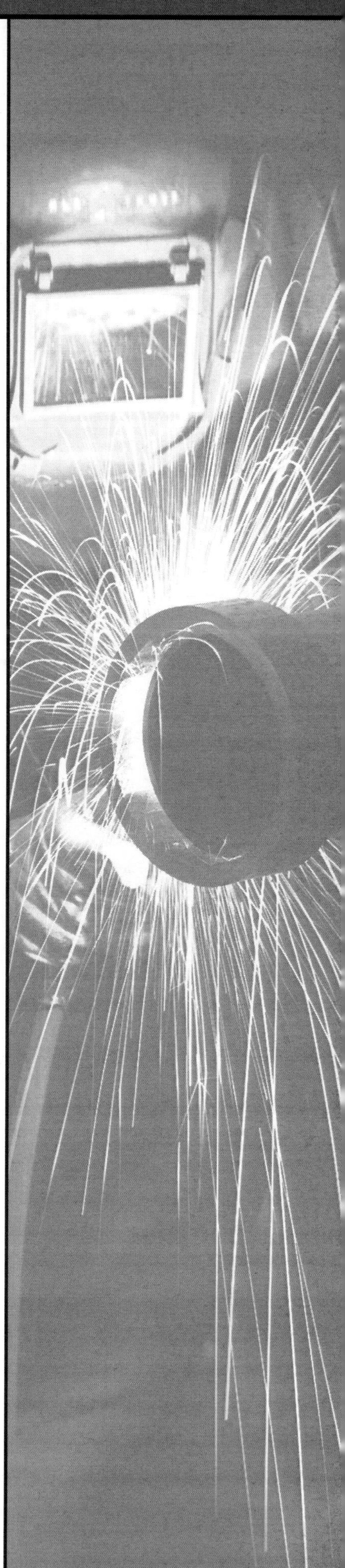

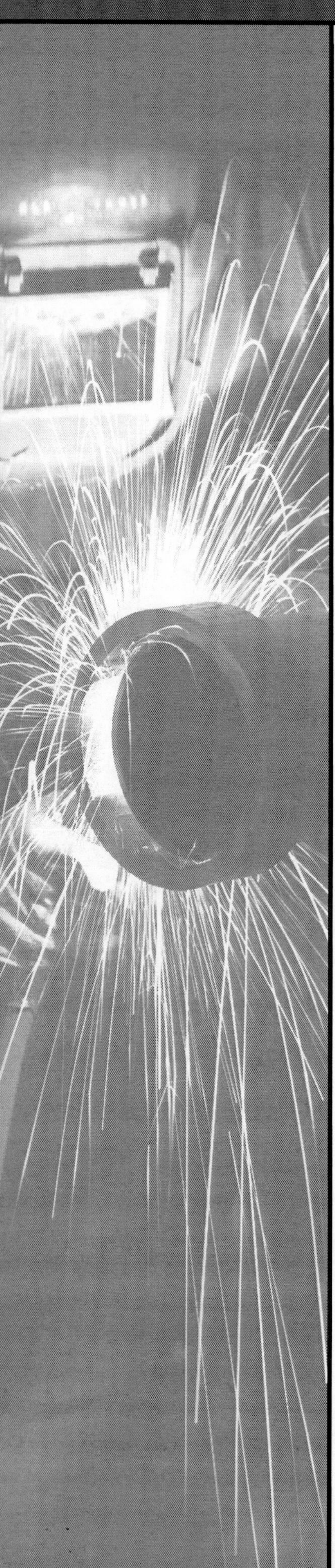

Contents

Figures and Tables

1.0.0 INTRODUCTION

Gas metal arc welding (GMAW) is an arc welding process that uses a continuous, consumable, solid wire electrode for the filler metal and a shielding gas to protect the weld zone (*Figure 1*). The GMAW process is commonly used to make welds on carbon steel, low-alloy steel, and stainless steel. It is also used for welds on aluminum and other metals. *Figure 2* shows typical GMAW equipment.

GMAW is a fast and effective method for producing high-quality welds. Because this type of welding can be continuous, discontinuities and restarts are reduced. With some materials, such as aluminum, it is a common field practice to use GTAW for the root pass and GMAW to complete the remaining passes. For the purposes of this training module, the dimensions used are representative of various codes and standards and may not be specific to any particular code. Always refer to the applicable code, standard, or site WPS.

WIRE ELECTRODE
CURRENT CONDUCTOR
SHIELDING GAS
WIRE GUIDE AND CONTACT TIP
ARC
WELD DIRECTION
SOLIDIFIED WELD MATERIAL
SHIELDING GAS
MOLTEN WELD METAL
401F01.EPS

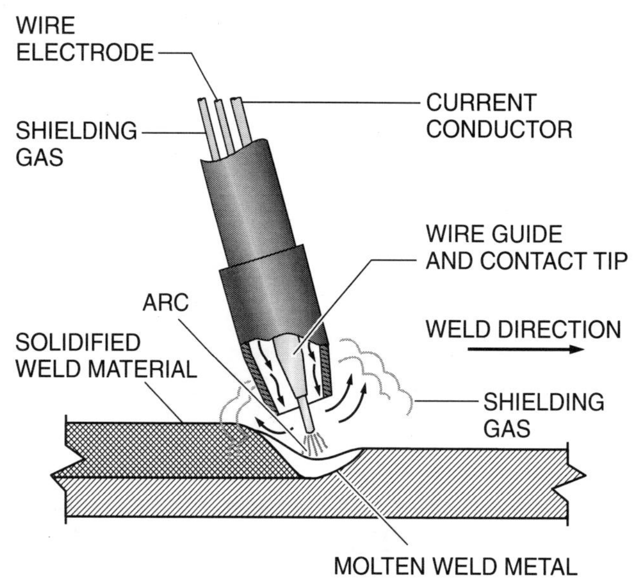

Figure 1 GMAW process.

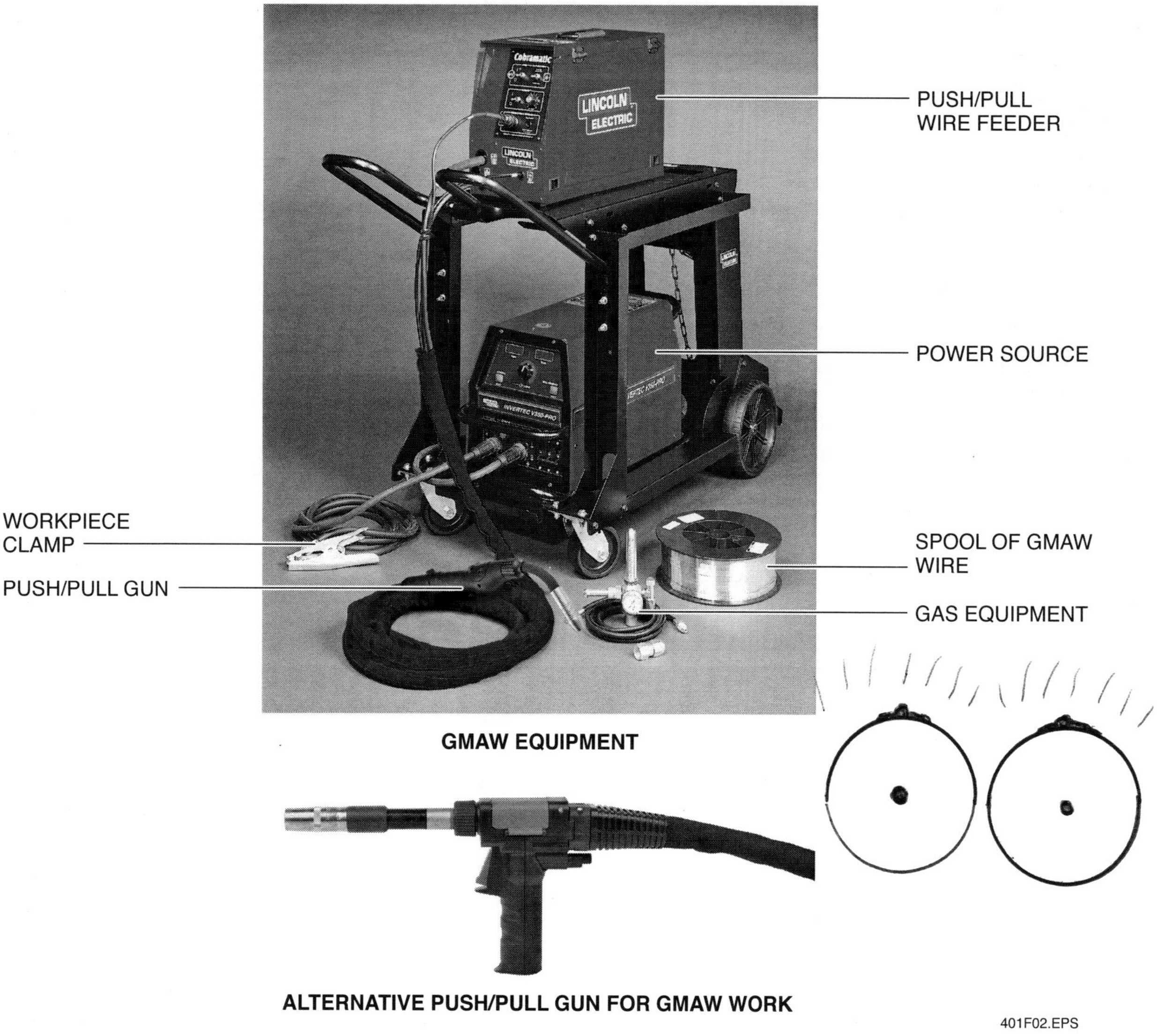

PUSH/PULL WIRE FEEDER
POWER SOURCE
WORKPIECE CLAMP
PUSH/PULL GUN
SPOOL OF GMAW WIRE
GAS EQUIPMENT
GMAW EQUIPMENT
ALTERNATIVE PUSH/PULL GUN FOR GMAW WORK
401F02.EPS

Figure 2 Gas-shielded GMAW equipment.

2.0.0 SAFETY SUMMARY

The following is a summary of safety procedures and practices that must be observed when welding. Keep in mind that this is only a summary; complete safety coverage is provided in the Level One module, *Welding Safety*. If you have not completed that module, do so before continuing. Above all, always use the appropriate protective clothing and equipment when welding.

2.1.0 Protective Clothing and Equipment

Welding work creates flying debris, such as sparks or small chunks of hot metal. Anyone welding or assisting a welder must use the proper protective clothing and equipment. The following list provides protective clothing and equipment guidelines:

- Always use safety glasses with a full face shield or a helmet. The glasses, face shield, or helmet lens must have the proper light-reducing tint for the type of welding being performed. Never directly or indirectly view an electric arc without using a properly tinted lens.
- Wear proper protective leather and/or flame retardant clothing and welding gloves. They will protect you from flying sparks, molten metal, and heat.
- Wear 8-inch or taller high-top safety shoes or boots. Make sure that the tongue and lace area is covered by a pant leg. Sometimes the tongue and lace area is exposed or the footwear must be protected from burn marks. In those cases, wear leather spats under the pants or chaps and over the front and top of the footwear.
- Wear a solid material (non-mesh) hat with a bill pointing to the back or toward the ear closest to the welding. This will give added protection. If much overhead welding is required, use a full leather hood with a welding faceplate and the correct tinted lens. If a hard hat is required, use one that allows attachment of both rear deflector material and a face shield.
- Wear earmuffs or earplugs to protect your ear canals from sparks.

2.2.0 Fire/Explosion Prevention

Welding work includes the cutting, grinding, and welding of metal. All of these actions generate heat and often produce flying sparks. The heat and flying sparks can be the cause of fires and explosions. Welders must use extreme care to protect both themselves and others near their work. The following list provides fire and explosion protection guidelines:

- Never carry matches or gas-filled lighters in your pockets. Sparks can cause the matches to ignite or the lighter to explode, causing serious injury.
- Never perform any type of heating, cutting, or welding until a hot-work permit has been obtained and an approved fire watch established. Most work-site fires in these types of operations are started by cutting torches.
- Never use oxygen to blow dust or dirt from clothing. The oxygen can remain trapped in the fabric for a time. If a spark hits the clothing during this period, the clothing can burn rapidly and violently out of control.
- Make sure that any flammable material in the work area is either moved to a safe area or shielded by a fire-resistant covering. Approved fire extinguishers must be available before attempting any heating, welding, or cutting operations.
- Never release a large amount of fuel gas, especially acetylene. Methane and propane tend to concentrate in and along low areas. Both gases can ignite at a considerable distance from the release point. Acetylene is lighter than air, but it is even more dangerous than methane. When mixed with air or oxygen, acetylene will explode at much lower concentrations than any other fuel.
- To prevent fires, maintain a neat and clean work area. Also, make sure that any metal scrap or slag is cold before disposal.

WARNING

Before welding containers, such as tanks or barrels, check to see if they have ever held any explosive, hazardous, or flammable materials. These include petroleum products, citrus products, or chemicals that decompose into toxic fumes when heated. As standard practice, always clean and then fill any tanks or barrels with water, or purge them with an appropriate purging gas to displace any oxygen.

GOING GREEN

Water Disposal

To protect the environment and save resources, make sure to properly dispose of any water used in the cutting of tanks, barrels, or metals. If the water can be reused, save it and use it again for the next cutting.

2.3.0 Work Area Ventilation

Welders normally work within inches of their welds, wearing special protective helmets. Vapors from the welds can be hazardous. The following list provides work area ventilation guidelines:

- Always follow the required confined space procedures before conducting any welding in the confined space.
- Never use oxygen for ventilation in confined spaces.

WARNING

An oxygen monitor may be required when working in a confined space.

- Always perform welding operations in a well-ventilated area. Welding operations involving zinc or cadmium materials or coatings result in toxic fumes. For long-term welding of these materials, always wear an approved, full-face SAR that uses breathing air supplied from outside of the work area. For occasional, very short-term exposure, you may use a HEPA-rated or metal-fume filter on a standard respirator.
- Make sure confined spaces are properly ventilated for welding operations.

3.0.0 Aluminum Metallurgy

Aluminum and aluminum alloys are often used in manufacturing. This is because of their low weight and good corrosion resistance. Aluminum is one-third the weight of steel. It has six times the thermal conductivity of steel. Pure aluminum is fairly soft, has low strength, and is not suitable for load carrying.

Aluminum is not toxic, nonsparking, and nonmagnetic. Molten aluminum has a silvery sheen under the glow of the arc. When heated, aluminum does not change color as steel does.

Aluminum is a highly reactive element. This explains why it is never found in nature in its free state, even though it is the most common metal on earth. Aluminum has a strong chemical inclination toward oxygen. This causes aluminum to oxidize as soon as it is exposed to air. Aluminum oxide is extremely hard and has a high melting point. It is also porous, allowing it to hold moisture and contaminants, which can result in weld porosity. This tendency is especially true of aluminum alloys containing magnesium.

Aluminum can be strengthened by the addition of small amounts of alloy metals to the base metal. Alloys are also used to enhance various other properties of aluminum. Alloys are grouped into series by their alloying elements. The alloys in each series are further defined by their ability to be treated by heat. Heat-treatable and nonheat-treatable aluminum alloys react differently to welding. Strain hardening is a method of increasing the strength of a metal by cold working it. Alloys that respond to cold working are said to be strain-hardenable. *Table 1* shows aluminum alloy characteristics by series, classification, and major alloying elements.

Aluminum metallurgy can be divided into the following categories:

- Nonheat-treatable alloys
- Heat-treatable alloys
- Filler metal alloys
- Cast aluminum
- Wrought aluminum

3.1.0 Nonheat-Treatable Alloys

The strength of nonheat-treatable aluminum alloys depends on elements such as silicon (Si), iron (Fe), manganese (Mn), and magnesium (Mg). These alloying elements are found in the 1xxx, 3xxx, and 5xxx series of alloys. When welded, these alloys may lose the effects of work hardening, which results in softening of the heat-affected zone (HAZ) next to the weld.

Additional strengthening comes from work hardening (strain hardening) in the forming process, such as by cold-rolling. Welds in nonheat-treatable alloys generally have excellent ductility and can withstand severe distortion before failing. Welds in heat-treatable alloys generally have lower ductility than welds in nonheat-treatable alloys. The basic temper designations for nonheat-treatable and heat-treatable aluminum alloys, such as castings, are as follows:

- *O* – Annealed, recrystallized
- *F* – As fabricated

Table 1 Aluminum Alloy Characteristics

Series	Classification	Major Alloying Elements	Characteristics
1xxx 1050A	Nonheat-treatable	High purity with small amounts of iron and silicon	Relatively soft, low-strength, nonload carrying; excellent corrosion resistance
2xxx 2014A	Heat-treatable	Copper	High strength; prone to liquation under welding heat
3xxx 3103	Nonheat-treatable	Manganese	Stronger and more ductile than 1xxx grades; excellent corrosion resistance
4xxx 4043A	Filler wires	Silicon	Excellent corrosion resistance
5xxx 5083	Nonheat-treatable	Magnesium	Good ductility; excellent corrosion resistance; good for situations involving impact and shock-loading; not recommended for service with prolonged exposure to high temperatures
6xxx 6082	Heat-treatable	Silicon and magnesium	Good as-welded strength; easily extruded; not recommended for exposure to electrolytes, such as seawater
7xxx 7020	Heat-treatable	Zinc	Good weldability and high as-welded strength
8xxx 8090	Nonheat-treatable and heat-treatable	Other	Not normally welded

401T01.EPS

- *H1(x)* – Strain hardened only, where (x) = 2, 4, 6, 8, or 9, representing ¼, ½, ¾, full hardness, and extra hardness
- *H2(x)* – Strain hardened and then partially annealed, where (x) = 2, 4, 6, 8, or 9, representing ¼, ½, ¾, full hardness, and extra hardness
- *H3(x)* – Strain hardened and then stabilized, where (x) = 2, 4, 6, 8, or 9, representing ¼, ½, ¾, full hardness, and extra hardness
- *W(HR)* – Solution heat treated, where (HR) = fractional and full hours of aging at room temperature after heat treatment
- *T(N)* – Thermally treated (other than O, F, or H), where (N) = 1 through 10, which indicates a specific treatment, sometimes followed by another digit representing a variation of the treatment

3.2.0 Heat-Treatable Alloys

Heat-treatable aluminum alloys develop their properties by solution heat treating and quenching, followed by either natural or artificial aging. Heat-treatable alloys contain copper, magnesium, zinc, or silicon. These elements become more soluble when heated. Welding heat reduces the strength of heat-treatable alloys, but longer times are required to modify their structure. As a result, factors such as welding speed and cooling rate have an influence on their final properties. Because of the time factor, the HAZ in heat-treatable alloys is usually only partially annealed. Fusion welding redistributes the hardening components in the HAZ. This locally reduces strength. The 2xxx, 6xxx, 7xxx, and some of the 8xxx series contain the principal heat-treatable alloying elements. The heat-treatable temper designations include O, F, W, and T1 through T10.

Postweld heat treatment increases the strength of the HAZ, but it reduces the amount of ductility. Strength improvement depends somewhat on the filler metal used. Preheating is seldom recommended for welding heat-treatable alloys.

Joint thickness can also have an effect on weld properties in heat-treatable alloys. Thinner metal can be welded with a minimum of heat input and with a high cooling rate. This results in higher as-welded strength. For thicker metal, a number of string beads are preferable to a few large beads.

3.3.0 Filler Metal Alloys

Filler metal composition is determined in part by the weldability of the base metal and the minimum properties of the weld metal. Other factors to consider include corrosion resistance and the **anodic coating** requirements. For nonheat-treatable alloys, matching fillers are often used.

For low-alloy materials and heat-treatable alloys, nonmatching fillers are used to prevent solidification cracking.

3.4.0 Cast Aluminum

Cast aluminum has high silicon content for increased fluidity. It does not anodize as well as wrought aluminum. Cast aluminum is identified by three-digit grade numbers, such as 206, 319, and 356. *Table 2* lists the numbers representing the alloying elements used with cast aluminum. The first number indicates the alloying element. The second and third numbers represent the specific alloy by formula. The number after the decimal point shows whether the alloy composition is for the final casting (0.0) or for the **ingot** (0.1 or 0.2, depending on purity). Modifications to cast aluminum are shown by a letter in front of the number. An example of a typical silicon-magnesium alloy for final casting is A356.0. The same alloy in ingot form would be A356.1 or A356.2.

3.5.0 Wrought Aluminum

Wrought aluminum alloys are identified by four numbers. *Table 3* shows the alloy series numbers and the principal alloying elements used with the various wrought aluminum alloys. The first number indicates the principal alloying element or elements. The second number, if not a zero, signifies some change to the alloy. The third and fourth numbers identify the specific alloy by formula. For example, alloy 5054 is an aluminum-magnesium alloy without modifications. It has the assigned alloy number of 54. Alloy 5154 is the same alloy after modification. The only exception is the 1xxx series, which is for pure aluminum. In that series, the last two numbers show the degree of purity above 99 percent. For example, alloy 1060 has a purity of 99.60 percent.

Most of the wrought grades in the 1xxx, 3xxx, 5xxx, 6xxx, and the medium-strength 7xxx series (for example, 7020) can be fusion-welded with GTAW and GMAW. The 5xxx series alloys have excellent weldability.

Wrought aluminum is available in five basic categories:

- Plate or sheet
- Rod, bar, or wire
- Tubing
- Extrusions
- Forgings

Table 2 Alloy Numbers for Cast Aluminum

Alloy Series	Principal Alloying Elements
1xx.x	Essentially pure aluminum
2xx.x	Copper
3xx.x	Silicon plus copper and/or magnesium
4xx.x	Silicon
5xx.x	Magnesium
6xx.x	Unused series
7xx.x	Zinc
8xx.x	Tin
9xx.x	Other elements

401T02.EPS

Table 3 Alloy Numbers for Wrought Aluminum

Alloy Series	Principal Alloying Elements
1xxx	Aluminum (99% minimum and greater)
2xxx	Copper
3xxx	Manganese
4xxx	Silicon
5xxx	Magnesium
6xxx	Magnesium and Silicon
7xxx	Zinc
8xxx	Other elements
9xxx	Unused series

401T03.EPS

4.0.0 Characteristics of Aluminum Welding

GMAW does not usually require the preheating of aluminum. However, the high thermal conductivity of aluminum (compared to steel) requires a high rate of heat input for fusion welding. In some cases, very thick sections of aluminum may require preheating (at less than 250°F). To get the necessary welding heat for aluminum, higher current, the addition of helium to the shielding gas mix, and more precise control of the welding variables are required for the GMAW of aluminum than for welding steel.

GMAW is done with direct current electrode positive (DCEP) constant voltage power sources. When welding thick aluminum that requires deeper penetration, a shielding gas of helium mixed with argon is sometimes used with the DCEP power source. Pulsed GMAW is used to reduce heat input and prevent burn-through in thin sections of aluminum.

Welding current is mainly related to the metal's thickness. This is because the current must be high enough to give the required penetration. If the weld puddle becomes too large, the welding speed may need to be increased beyond the speed that is comfortable for the welder. To prevent melt-through on thin materials and on root passes with complete penetration, backing strips or rings made of stainless steel or a nonmetallic material are sometimes used. The backing strips or rings are often made with a shallow groove in

Hot Tip

Thickness Versus Required Current

The following is a guideline for estimating the amperage for welding aluminum:

For each 0.001" of metal thickness under the weld, a welding current of about one amp is required.

the surface. The groove ensures that penetrating welds have the required size and shape when cool. On advanced welding machines used for GMAW, pulsing control can be used to help eliminate melt-through problems.

4.1.0 Surface Preparation

Aluminum picks up hydrogen gas from condensation which forms during a rapid rise in temperature. Because materials are often stored in unheated areas, special care should be taken with aluminum to raise its temperature slowly. Another recommendation is to store aluminum in the vertical position to allow the runoff of any condensation. To reduce the risk of condensate, allow aluminum to remain at room temperature for one hour before welding.

Aluminum combines with oxygen from the atmosphere to form a surface film of very hard oxide. This process can even occur at room temperature. Pure aluminum melts at 1,220°F, but its oxide melts at 3,600°F to 3,900°F. Aluminum oxide is harder than most of the substances it encounters. It can destroy a milling cutter if the oxides are trapped in the welds. Because of its hardness, aluminum oxide is often used for the grit in grinding wheels. Aluminum oxide also tends to absorb moisture. Under the extreme heat of welding, the moisture breaks down to free hydrogen, which often leads to porosity in the weld. As a safeguard against porosity, oxidation must be removed from the base metal before welding.

WARNING

Grinding wheels used on aluminum must be made for aluminum only. They should not be used on any other type of metal or on surfaces contaminated with grease or oil. Using other types of grinding wheels on aluminum may cause the grinding wheels to shatter.

Water, oil, and grease are contaminants often found on aluminum. Wire brushing only smears the contaminants into the surface or damages the aluminum. Oil and grease must be chemically removed before cleaning aluminum with a wire brush.

4.1.1 Water-Based Cleaners

Both water-based aluminum deoxidizers and solvent cleaners (such as acetone, mineral spirits, and denatured alcohol) can also be used to remove grease. When using water-based cleaners, all water must be removed before welding. Most hydrocarbon-based solvents are highly volatile and evaporate quickly. Water-based cleaners must be thoroughly wiped away or heat-dried before welding. Review the MSDS for any additional safety hazards or precautions associated with these cleaners.

4.1.2 Chemical Cleaners

Preweld cleaning solutions can remove both the aluminum oxide and any contaminants. Their disadvantage is that deoxidizers for aluminum are very aggressive chemicals. They can be dangerous to the people using them. Always follow the recommendations in the MSDS when using cleaning solvents. Preweld cleaning products usually contain dilute acids. It is very important to follow the manufacturer's instructions and take the necessary safety precautions. When using etchants or deoxidizers, the metal must be completely rinsed and dried. This avoids creating a surface film of hydrated oxide.

Hot Tip

Clean, Clean, and Clean Again

Proper cleaning of welding surfaces cannot be stressed enough. Oxides, greases, and oil films contain oxygen and hydrogen. They can act as contaminants and cause unsound welds. Aluminum should be welded as soon as possible after cleaning or within one-half hour of cleaning. Otherwise the weld may have to be cleaned again. If oxides are not removed to the fullest extent possible, the weld will be subject to poor fusion and inclusions. This will be due to the reduced wetting action of the molten weld pool as it comes in contact with any remaining oxide-contaminated weld joint surfaces. Aluminum should be hand-brushed between passes with a stainless steel brush. Make sure that the brush has not been used on other materials or for any other purpose.

> **WARNING**
>
> Be very careful when using preweld cleaning solutions. They can harm not only the environment, but also the people using them. Make sure that all cleaning solutions are properly contained. Always be sure to wear the appropriate PPE. Refer to the applicable MSDS for instructions on how to properly dispose of used cleaning solutions.

4.2.0 Weld Problems

Before fusion welding can take place, any oxide must be removed. The melting point of the oxide is about three times that of aluminum. Because of this, the welding heat needed to break down the oxide causes the aluminum to melt long before the oxide does, and **coalescence** cannot take place. Removing the oxide prior to welding helps to avoid this problem.

The following are the most common problems in aluminum welding:

- Porosity
- Cracking
- Poor weld bead profile

4.2.1 Porosity

Aluminum welds are highly prone to porosity. The main cause of porosity is the absorption of hydrogen into the weld pool. The hydrogen forms separate pores in the weld metal as the weld pool solidifies. Some common sources of hydrogen are hydrocarbons and the atmosphere. Another is the moisture from contaminants on the parent material and on the surface of the filler wire. In some cases, a torch gas flow that is set too high can cause porosity in the weld. It does this by creating turbulence that draws air into the gas, causing porosity. *Figure 3* shows porosity caused by hydrocarbon contamination in a fillet weld.

Thoroughly clean the surface to reduce the risk of porosity. Proper cleaning techniques include solvent degreasing and chemical etching. These are followed by mechanical cleaning using methods such as grinder buffing or wire brushing. *Table 4* shows the causes, contributing factors, and corrective measures related to porosity and cracking in GMAW.

4.2.2 Weld Cracking in General

Cracking occurs in aluminum alloys because of high stress. The stresses are generated across the weld by the high degree thermal expansion, which is twice that of steel. Thermal expansion is the fractional increase in size due to a rise in temperature. Cracking can also be traced to the amount of contraction that occurs when the weld solidifies. Contraction is usually 5 percent more in aluminum than in similar welds on steel. High-strength aluminum alloys (such as 7010 and 7050) and most of the 2xxx series are not recommended for fusion welding. This is because they are prone to both solidification and **liquation** cracking.

Figure 3 Porosity example.

Table 4 Causes and Cures for Porosity and Cracking in GMAW

Cause	Contributing Factors	Corrective Measures
Hydrogen	Hydrogen is the result of dirt-containing oils or other hydrocarbons; moisture in atmosphere or on metal; a hydrated oxide film on metal; or moisture in gas or gas lines. Base metal may be the source of entrapped hydrogen (the thicker the metal, the greater the possibility of hydrogen). Spatter contributes to hydrogen formation.	Degrease and mechanically or chemically remove oxide from weld area. Avoid humidity; use dry metal or wipe dry. Reduce moisture content of gas. Check gas and water lines for leaks. Increase gas flow to compensate for increased hydrogen in thicker sections. To minimize spatter, adjust welding conditions.
Impurities	Cleaning or other compounds, especially those containing calcium.	Use recommended cleaning compounds; keep work free of contaminants.
Incomplete root penetration	Insufficient welding current or preheat.	Use a higher welding current, or redesign the joint geometry. Preheat.
Temperature	Running too cool tends to increase porosity that is caused by premature solidification of molten metal.	Maintain the proper current, arc-length, and torch-travel speed relationship.
Welding speed	Too great a welding speed may increase porosity.	Decrease the welding speed and establish and maintain proper arc-length and current relationship.
Solidification time	Quick freezing of the weld pool entraps any gases present, causing porosity.	Establish the correct welding current and speed. If the work is appreciably below room temperature, use supplemental heating.
Chemical composition of weld material	Pure aluminum weld metal is more susceptible to porosity than is an aluminum alloy.	If porosity is excessive, try an alloy filler material.
Cracking	Such causes of porosity as temperature, welding time, and solidification may also be contributing causes of cracking. Other causes may be discontinuous welds, welds that intersect, repair welds, cold-working either before or after welding, and weld-metal composition. In general, crack-sensitive alloys include those containing 0.4 to 0.6% silicon, 1.5 to 3.0% magnesium, or 1.0% copper.	Lower current and faster speeds often prevent cracking. However, a change to a filler alloy that brings the weld-metal composition out of the cracking range is recommended where possible.

401T04.EPS

4.2.3 Solidification Cracking

Solidification cracks (also called hot cracks) form in the center of the weld and run along its centerline. Solidification cracks also occur in the weld crater at the end of the weld. There are two main causes of solidification cracks. One is the incorrect combination of filler wire and base metal. The other is welding under high-restraint conditions.

Use of a non-matching, crack-resistant filler (usually from the 4xxx and 5xxx series alloys) can reduce the risk of cracking. However, the use of these fillers can result in a weld of lower strength. Also, the weld may not respond to heat treatment. The weld bead must be thick enough to withstand the stresses caused by contraction as the weld solidifies. There are several ways to reduce the degree of restraint on the weld. Methods include correct edge preparation, accurate joint setup, more compatible filler alloy, and correct weld sequence. Higher weld speeds can also reduce solidification cracking.

Preheating reduces stress on the solidifying weld metal. It does this by reducing the temperature change across the weld zone and by permitting faster joining. However, to reduce the risk of stress cracking, preheating should only be used when the joint is unrestrained. Preheating should be limited to temperatures under 250°F to preserve the mechanical properties of the base metal.

4.2.4 Liquation Cracking

Liquation cracking occurs in the HAZ at the grain boundaries when low-melting-point films are formed. These films are not strong enough to withstand the contraction stresses created when the weld metal solidifies and cools. Heat-treatable alloys, such as the 6xxx, 7xxx, and 8xxx series, are more likely than other alloys to crack in this way.

Use of a filler metal that has a lower melting temperature than the base metal can reduce the risk of cracking. An example would be to use a 4xxx filler metal to weld a 6xxx-series-alloy base metal. However, 4xxx filler metal should not be used to weld high-magnesium alloys, such as 5083. This is because excessive magnesium silicide may form at the fusion boundary, causing a loss of ductility and an increased risk of cracking.

4.2.5 Poor Weld Bead Profile

A poor weld bead profile can result from setting incorrect welding parameters. It can also be traced to poor welding techniques. Some common weld profile defects are lack of fusion, lack of penetration, and undercut. Aluminum alloys are especially prone to profile imperfections. This tendency is due to their high thermal conductivity and the rapidly solidifying weld pool.

5.0.0 WELDING PREPARATION

Before welding can begin, the area must be readied, the welding equipment set up, and the metal to be welded prepared. The following sections explain how to set up the equipment for welding.

To practice welding, you will need a welding table, bench, or stand. The welding surface can be steel, but an aluminum surface is preferred. Provisions must be available for placing weld coupons out of position.

To set up the area for welding, follow these steps:

Step 1 Make sure that the area is properly ventilated. Make use of doors, windows, and fans.

Step 2 Check the area for fire hazards. Remove any flammable materials before proceeding.

Step 3 Locate the nearest fire extinguisher. Do not proceed unless the extinguisher is charged and you know how to use it.

Step 4 Set up flash shields around the welding area.

5.1.0 Practice Welding Coupons

When possible, cut welding coupons from ⅜" thick aluminum plate. If this size is not readily available, you can use aluminum plate that is ¼" to ¾" thick.

Aluminum is very prone to contamination. Careful coupon preparation is important. First use an approved solvent, such as acetone, to remove oil and grease from the coupons. Then use a stainless steel wire brush to remove all other contaminants and oxidation. Be sure to use only light pressure.

To prevent contamination, do not use a stainless steel wire brush on anything else after it has been used on aluminum.

To avoid injury, always follow the MSDS guidelines for the cleaning solvents being used.

5.1.1 Plate Coupons

The coupons must be shaped to allow the following welds:

- *Stringer or weave beads (both running and overlapping)* – The coupons can be any size or shape that is easily handled.
- *Fillet welds* – Cut the metal into 4" × 5" rectangles for the base and 3" × 5" rectangles for the web (*Figure 4*).
- *V-groove welds with backing* – Refer to joint G of *Figure 5* and to *Figure 6* for the following instructions. Cut the metal into 3" × 5" rectangles with one or both of the 5" lengths beveled at 30 degrees. Grind a $\frac{1}{32}$" to ⅛" root face on the bevel as directed by your instructor.

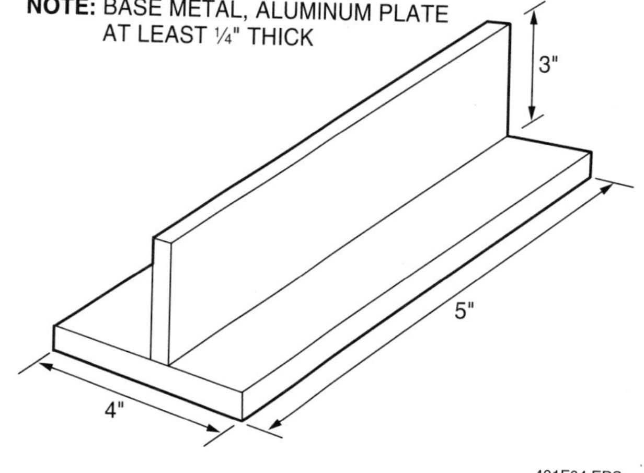

Figure 4 Fillet weld coupons.

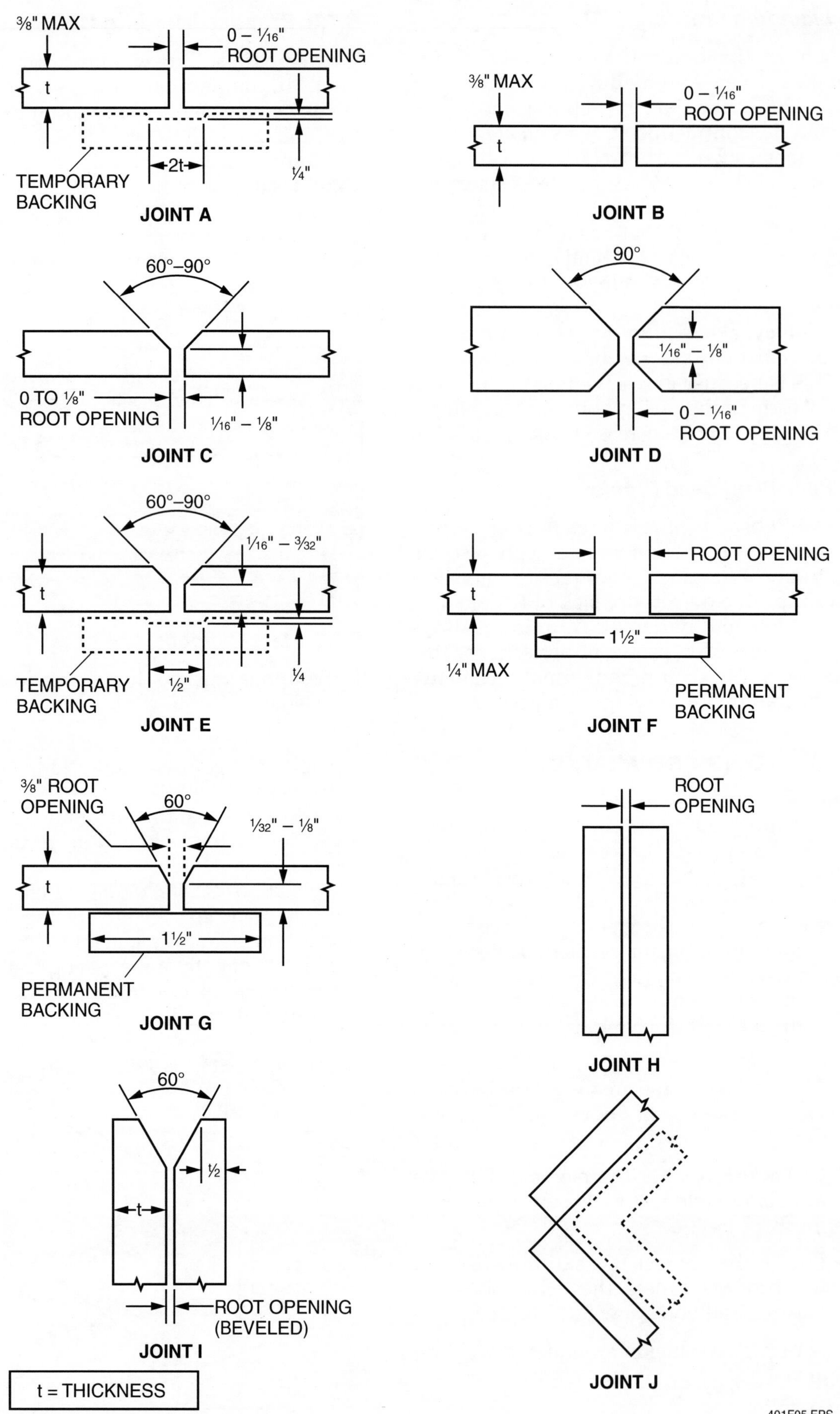

Figure 5 GMAW joint preparations and geometrics.

NOTE

This module uses joint G in *Figure 5* for practice purposes. However, the other configurations can also be used at the direction of the instructor.

5.1.2 Backing Strips

Cut backing strips at least 1½" wide and 6" long from the same metal as the beveled pieces. *Figure 7* shows two prepared V-groove coupons with a metal backing weld coupon.

Follow these steps to tack-weld each V-groove weld coupon with backing:

Step 1 Check the bevel angle. It should be about 30 degrees. The root face should be ⅛".

Step 2 Center the beveled strips on the backing strip with a ⅜" root opening (*Figure 7*), and tack-weld them in place. If desired, use a piece of metal or rod of the required thickness as a temporary spacer. Place the tack welds on the reverse side of the joint in the lap formed by the backing strip and the beveled plate. Use three to four ½" tack welds on each beveled plate. Ensure that the backing strip is tight against the beveled plates.

NOTE

Make sure that the surface of the backing strip under the root opening has been cleaned of all oxide coating before you tack the strip in place.

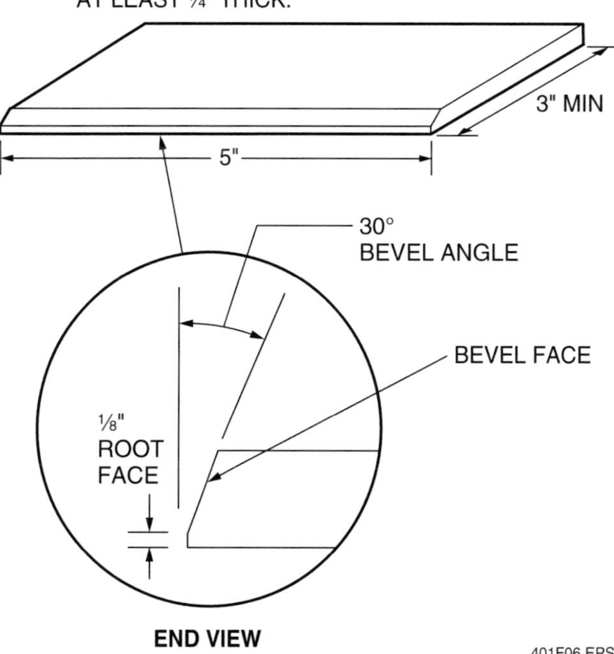

Figure 6 Weld coupon dimensions.

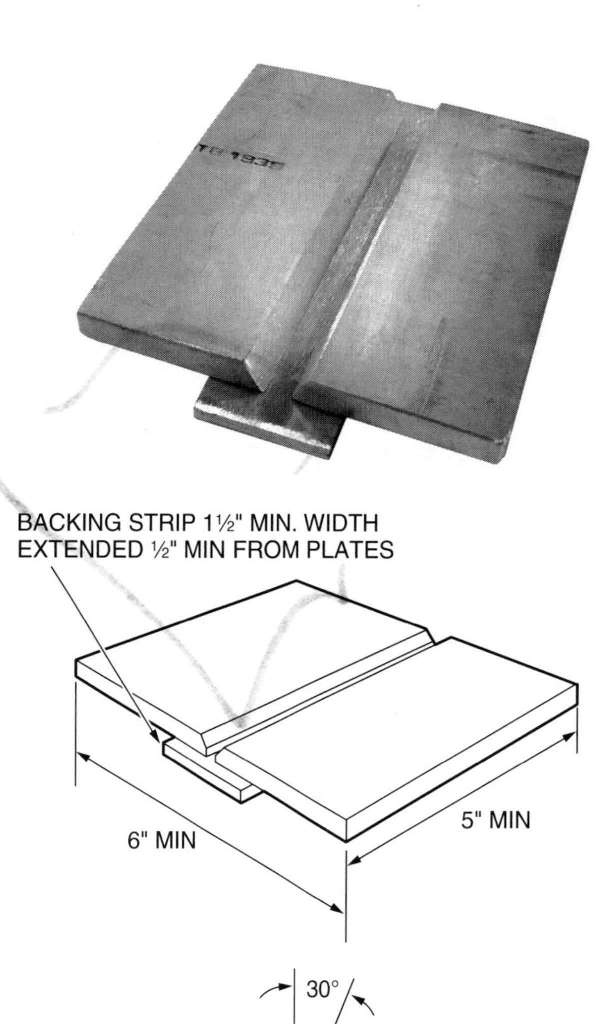

Figure 7 V-groove with metal backing weld coupon.

GOING GREEN

Conserve Aluminum Practice Coupons

For practice welding, aluminum is expensive and difficult to obtain. To conserve resources and reduce waste in landfills, completely use weld coupons until all surfaces have been welded upon. Cutting coupons apart and reusing the pieces conserves materials. Use material that cannot be cut into weld coupons to practice running beads.

> **Hot Tip**
>
> **Backing Strip**
>
> When the backing strip is ½" to 1" longer than the plates at each end (*Figure 7*), it allows you to start and stop the bead outside of the weld groove.

5.1.3 Alternate Joint Preparation

An alternate joint preparation can be used when welding in the horizontal position. This variation has one plate beveled at about 45 degrees and the other at about 15 degrees. The plate beveled at 45 degrees is placed above the 15-degree plate with a ⅜" root opening (*Figure 8*). Note that the degrees shown are approximate.

> **NOTE**
>
> Check with your instructor about whether to use the standard V-groove preparation or the alternate preparation for horizontal weld coupons.

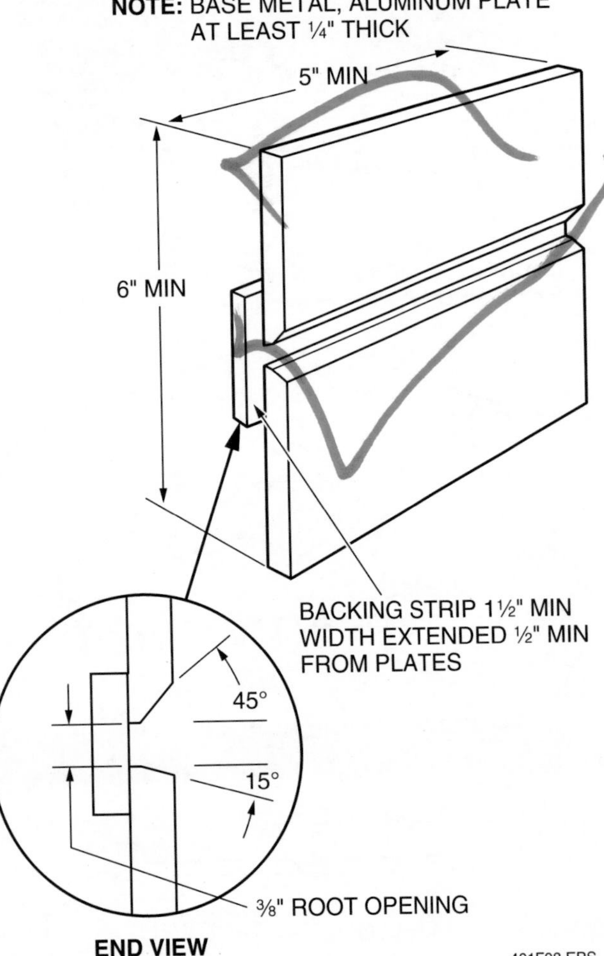

Figure 8 Alternate horizontal weld coupon.

> **Hot Tip**
>
> **Grinding Wheels**
>
> Some WPS or site quality standards may prohibit grinding aluminum welds. Always check before using a grinding wheel. When using a grinding wheel, make sure that it is one approved for use on aluminum. Also, make sure that it has not been used to grind a hydrocarbon-contaminated surface or any other metals.

5.2.0 The Welding Machine

Identify a proper welding machine for GMAW. Follow these steps to set the machine up for use:

Step 1 Verify that the welding machine can be used for GMAW.

Step 2 Know the location of the primary disconnect.

Step 3 Configure the welding machine for GMAW (*Figure 9*) as directed by your instructor. Configure the gun polarity (DCEP). Equip the gun with the correct nozzle for the application and the proper liner material and contact tube for the diameter of wire being used.

> **NOTE**
>
> The pull system shown in *Figure 9* is the most commonly used method. Spool guns are another option.

Step 4 In accordance with the manufacturer's instructions, configure and load the wire feeder and gun with the proper type and diameter of electrode wire.

> **CAUTION**
>
> To prevent the electrode wire from contaminating the welds, always keep the wire clean when it is in storage or in the machine.

Step 5 Connect the proper shielding gas (argon or argon/helium) for the application as described in the previous level and as specified by the wire electrode manufacturer, WPS, site quality standards, or your instructor.

Step 6 Connect the clamp of the workpiece lead to the workpiece.

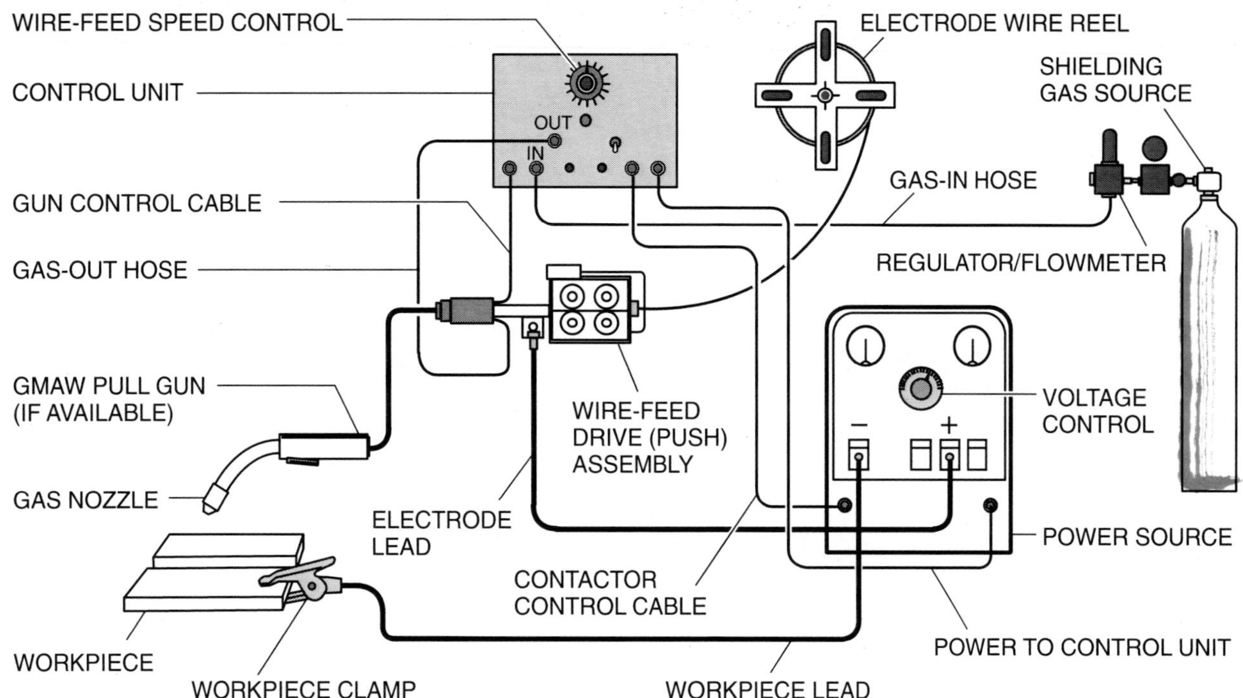

Figure 9 Configuration diagram of a typical GMAW machine.

Step 7 Turn on the welding machine. Purge the gun as directed by the gun manufacturer's instructions.

Step 8 Set the initial welding voltage and wire feed speed as recommended by the manufacturer for the type and size of electrode wire being used.

5.2.1 Voltage

Arc length is determined by the voltage, which is set at the power source. Arc length is the distance from the wire electrode tip to the base metal or to the molten pool at the base metal (*Figure 10*). If voltage is set too high, the arc will be too long. This can cause the wire to melt and fuse to the contact tube. Voltage that is too high can also cause porosity and excessive spatter. Voltage must be increased or decreased as the wire feed speed is increased or decreased. Set the voltage to maintain consistent spray transfer.

> **Hot Tip**
>
> **Contact Tube Size for Aluminum Wire**
>
> The standard practice is to use a contact tube one size larger than the diameter of the wire.

5.2.2 Amperage

When using a standard constant-voltage power source, the electrode feed speed controls the welding amperage, after the initial recommended setting. The welding power source provides the

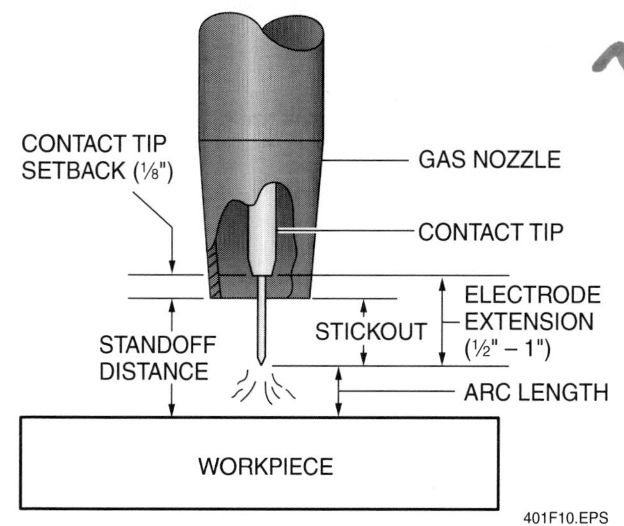

Figure 10 Arc length.

Hot Tip

Wire Electrode Manufacturer's Recommendations

Always obtain and follow the manufacturer's recommendations for shielding gas and for the initial setting of the welding voltage and wire feed speed parameters. These balanced parameters are critical and are based on the welding position, size, and composition of the solid wire.

Arc Voltage

A minimum arc voltage is needed to maintain spray transfer. However, penetration is not directly related to voltage. Penetration will increase with voltage for a time, but it will actually decrease if the voltage is increased above its optimum.

amperage needed to melt the wire electrode while maintaining the selected welding voltage. Within limits, when the wire electrode feed speed is increased, the welding amperage and deposition rate are also increased. This results in higher welding heat, deeper penetration, and higher beads. When the wire electrode feed speed is decreased, the welding amperage automatically decreases. With lower welding amperage and less heat, the deposition rate drops, and the weld beads are smaller with less penetration.

Note that some constant-voltage power sources used for GMAW/FCAW provide varying degrees of current modification. Examples include slope and induction adjustments. Power sources with a slope adjustment allow the welder to vary the amount of amperage change in relation to the voltage range of the unit. Standard constant-voltage GMAW/FCAW units have a current slope fixed by the manufacturer for general welding applications and conditions. Pulse transfer power sources allow the peak current for the pulse and the background current between pulses to be adjusted to match specific welding applications and/or wire electrode requirements.

5.2.3 Weld Travel Speed

Weld travel speed is the speed at which the electrode tip passes across the base metal in the direction of the weld. It is measured in inches per minute (ipm). Travel speed has a great effect on penetration and bead size. Slower travel speeds build higher beads with deeper penetration.

Hot Tip

Travel Speed and Wire Feed Speed

New welders are tempted to turn down the wire feed speed if they have difficulty controlling the weld puddle. Wire electrodes must be run at certain balanced parameters that cannot be changed individually. Voltage, current (if variable), wire feed speed, and travel speed are adjusted and balanced together to control the weld puddle.

Faster travel speeds build smaller beads with less penetration. Ideally, the welding parameters should be adjusted so that the electrode tip is positioned at the leading edge of the weld puddle during travel.

5.2.4 Gun Position

The gun position in relation to the direction of the weld is defined for aluminum as shown in *Figure 11*. A push angle is used to allow base metal cleaning.

5.2.5 Electrode Extension, Stickout, and Standoff Distance

Electrode extension is the length of the wire that extends beyond the tip of the welding gun's contact tip. For high-conductivity metal wires, the preheating effect of wire resistance is minimal. Wire speed and voltage settings have a more direct effect on weld penetration and deposition rate.

GMAW electrode extension for spray transfer varies from ½" to 1". *Figure 12* shows the typical electrode extension for the spray transfer GMAW gun configuration. It also shows various gun terms and components. Stickout is the distance from the gas nozzle or insulating nozzle to the end of the electrode. Standoff distance is the distance from the gas nozzle or insulating nozzle to the workpiece. Contact tip setback is important in the spray or transfer mode of GMAW. The setback protects the contact tip from the additional heat created during the spray or pulse transfer mode. Reducing the heat helps to keep the wire from melting in the contact tip.

5.2.6 Gas Nozzle Cleaning

As the welding machine is being used, weld spatter builds up on the gas nozzle, diffuser, and contact tube. The gas nozzle must be cleaned from

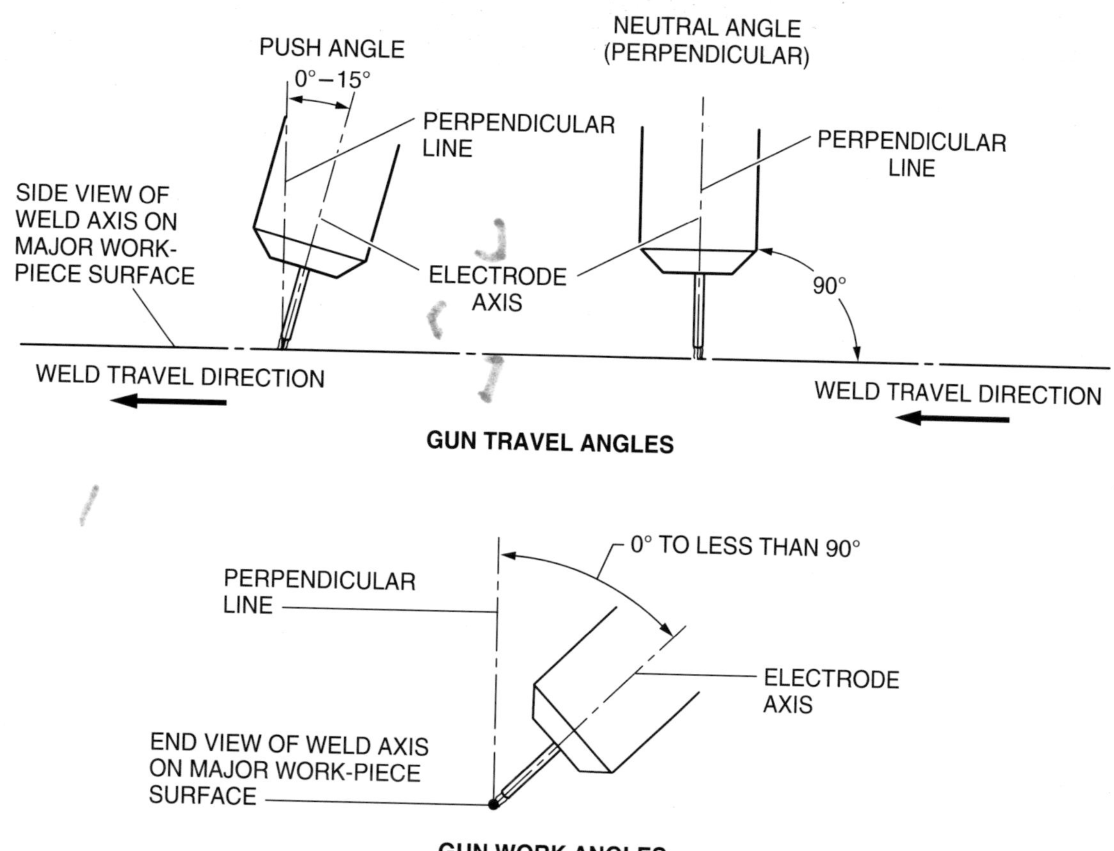

Figure 11 Gun work and travel angles.

time to time with a reamer, round file, or the tang of a standard file. If it is not properly cleaned, it will restrict the shielding gas flow. This will cause porosity in the weld.

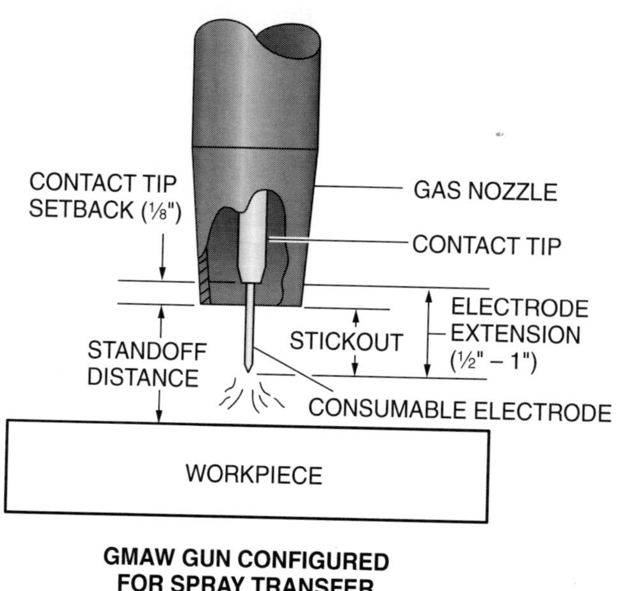

Figure 12 GMAW gun configuration for aluminum welding.

6.0.0 Welding Beads

The two basic bead types are stringer beads and weave beads. Aluminum tends to get contaminated more easily than steels. Because of this, extreme care must be taken to clean the base metals just before starting either type of bead. Always refer to the WPS for cleaning instructions and for the proper chemicals to use to remove oil and grease.

6.1.0 Bead Types

Stringer beads (*Figure 13*) are made with little or no side-to-side motion of the gun. Practice running stringer beads in the flat position. Experiment with different push angles and stickouts.

Follow these steps to make stringer beads:

Step 1 Clean the base metal before welding.

WARNING

To prevent personal injury when using cleaning solvents, always comply with the MSDS guidelines for the cleaning solvents being used.

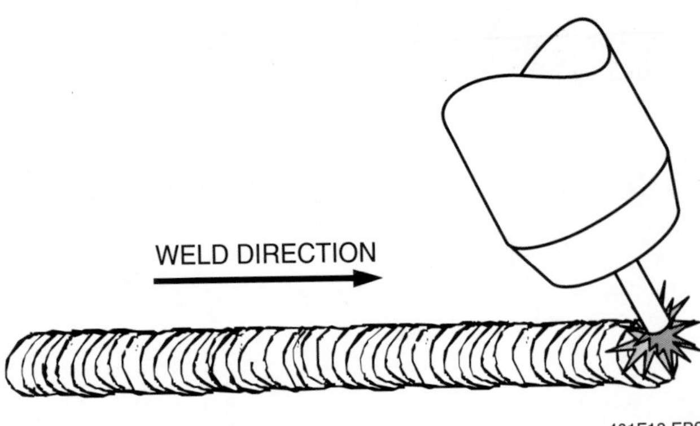

Figure 13 Stringer bead.

Step 2 Hold the gun at the desired angle with the electrode tip directly over the point where the weld will begin. Pull the gun trigger.

Step 3 Hold the arc in place until the weld puddle begins to form.

Step 4 Slowly advance the arc while maintaining the gun angle.

> **CAUTION**
> Stay in the leading edge of the puddle to maintain cleaning action and to prevent lack of fusion.

Step 5 Stop forward movement and back up about ⅛".

Step 6 Release the trigger once the crater has been filled.

Step 7 Inspect the bead for the following:
- Straightness
- Uniform appearance
- Smooth, flat transition with complete fusion at the toes of the weld
- Crater filled
- No porosity
- No excessive undercut
- No cracks

Step 8 Continue practicing stringer beads until you can make acceptable welds every time.

Weave beads (*Figure 14*) are made with wide side-to-side motions of the electrode called oscillations. The width of a weave bead is determined by the amount of side-to-side motion.

When making a weave bead, take special care at the toes to ensure proper fusion to the base metal. Do this by slowing down or pausing slightly at the edges. Pausing at the edges will also flatten out the weld and give it the proper profile.

> **CAUTION**
> Do not exceed the recommended weave bead width, which is often specified in the welding code or WPS used at your site.

Practice running weave beads in the flat position. Experiment with different weave motions, push angles, and stickouts.

Follow these steps to make weave beads:

Step 1 Clean the base metal before welding.

Step 2 Hold the gun at the desired angle with the electrode tip directly over the point where the weld will begin. Pull the gun trigger.

Step 3 Hold the arc in place until the weld puddle begins to form.

Step 4 Slowly advance the arc in a weaving motion (*Figure 15*) while maintaining the gun angle.

Step 5 Finish the bead by pausing until the crater has been filled, and then release the trigger.

Figure 14 Weave bead.

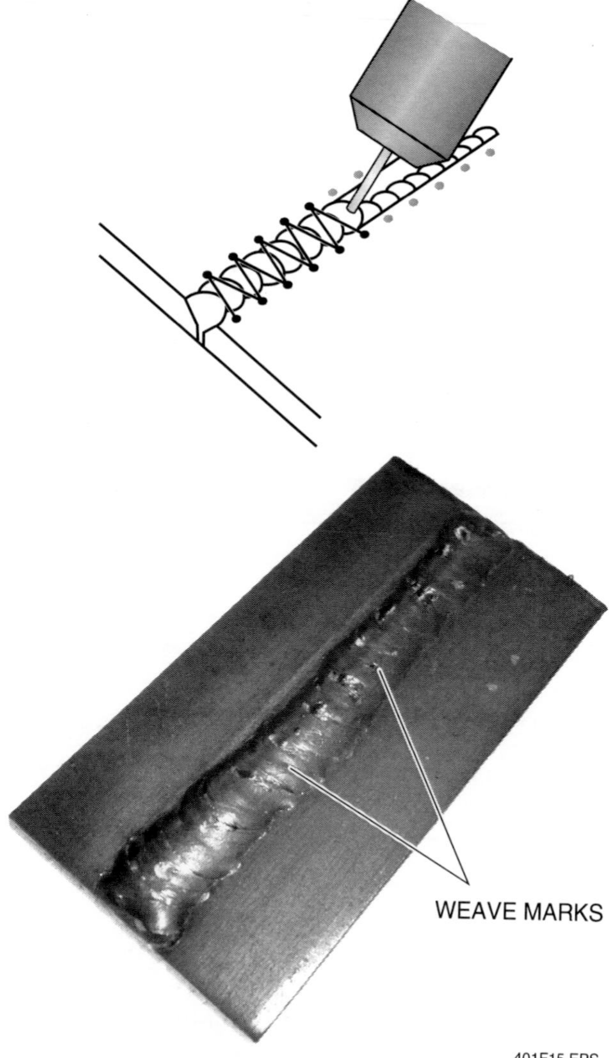

Figure 15 Weave motion.

Step 6 Inspect the bead for the following:

- Straightness
- Uniform appearance
- Smooth, flat transition with complete fusion at the toes of the weld
- Crater filled
- No porosity
- No excessive undercut
- No cracks

Step 7 Continue practicing weave beads until you can make acceptable welds every time. Clean the beads between each pass.

6.2.0 Weld Restarts

A restart is the junction where a new weld connects to and continues the bead of a previous weld. A restart must blend smoothly with the previous weld and must not stand out or create a weld defect. The technique for making a GMAW restart is the same for both stringer and weave beads.

Follow these steps to make a restart:

Step 1 Clean the base metal before welding.

Step 2 Hold the gun at the proper angle while restarting the arc directly over the center of the crater.

> **NOTE**
> Welding codes do not allow arc strikes outside of the area that is to be welded.

Step 3 Move the electrode tip in a small circular motion over the crater to fill it with a molten puddle.

Step 4 Move to the leading edge of the puddle. As soon as the puddle fills the crater, continue the stringer or weave bead pattern.

Step 5 Inspect the restart.

> **NOTE**
> A properly made restart blends into the bead and is hard to detect. If the restart has undercut, not enough time was spent in the crater to fill it. If the undercut is on one side or the other, use more of a side-to-side motion as you move back into the crater. If the restart has a lump, it was overfilled. Too much time was spent in the crater before the forward motion was resumed.

Step 6 Continue to practice restarts until they are correct.

> **NOTE**
> Use the same technique for making restarts whenever performing GMAW.

6.3.0 Weld Terminations

A weld termination normally leaves a crater. Most welding codes require the crater be filled to the full cross section of the weld when making a termination. This can be difficult because most terminations are at the edge of a plate where welding heat tends to build up. This makes it harder to fill the crater.

To terminate a weld, see the steps shown in *Figure 16* and those described in the following procedure:

Step 1 Start to bring the gun up to a 0-degree travel angle, and slow the forward travel as you approach the end of the weld.

Step 2 Stop forward movement and back up about ⅛".

Step 3 Release the trigger when the crater has been filled.

CAUTION

Do not remove the gun from the weld until the puddle has solidified. The shielding gas post-flow that continues after the welding has stopped protects the molten aluminum. If the gun and shielding gas are removed before the crater has solidified, crater porosity or cracks can occur.

Step 4 Inspect the termination. The crater should be filled to the full cross section of the weld.

6.4.0 Overlapping Beads

Overlapping beads are made by depositing connective weld beads parallel to one another. The parallel beads overlap to form a flat surface. This is also called padding. Overlapping beads are used to build up a surface and to make multiple-pass welds. Both stringer and weave beads can be overlapped. Properly overlapped beads, when viewed from the end, form a relatively flat surface. *Figure 17* shows the difference between proper and improper overlapping beads.

Follow these steps to weld overlapping stringer or weave beads using aluminum wire electrodes and the appropriate shielding gas:

Hot Tip

Alternate Crater Filling or Elimination Techniques

One way to eliminate the crater when terminating a GMAW weld on aluminum is to use run-off tabs. Sometimes tabs cannot be used or are impractical, such as with pipe welds. In those cases, you can reduce the weld pool before extinguishing the arc by increasing the travel speed just before releasing the gun trigger. This results in a small crater that may be free of cracks. Another technique for filling the crater without adding heat to increase its size is to make several quick arc starts into the crater as it cools.

Step 1 Mark out a square on a piece of aluminum.

Step 2 Clean and deoxidize the base metal.

Step 3 Weld a stringer or weave bead along one edge.

Step 4 Clean the bead with a stainless steel brush.

Step 5 Position the gun at a work angle of 10 to 15 degrees toward the side of the previous bead to get proper fusion. Pull the trigger.

Step 6 Keep running overlapping stringer or weave beads until the square has been covered.

Step 7 Continue building layers of stringer or weave beads, one on top of the other, until your technique has been perfected.

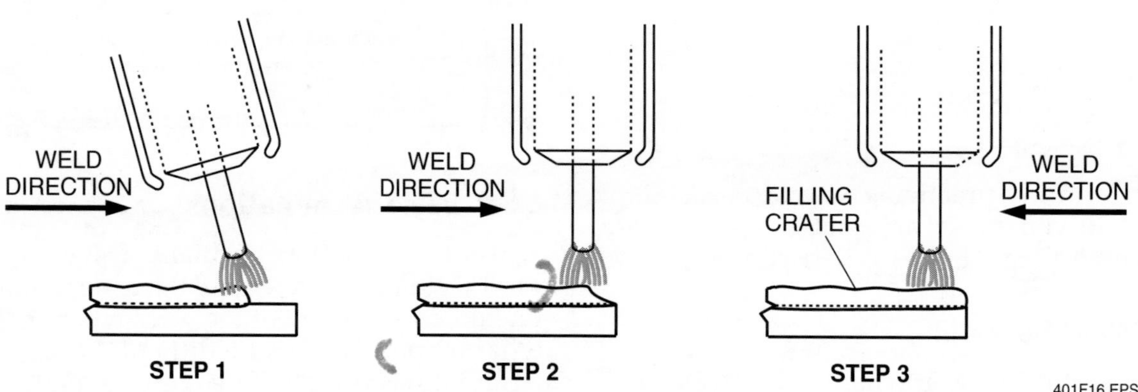

Figure 16 Terminating a weld.

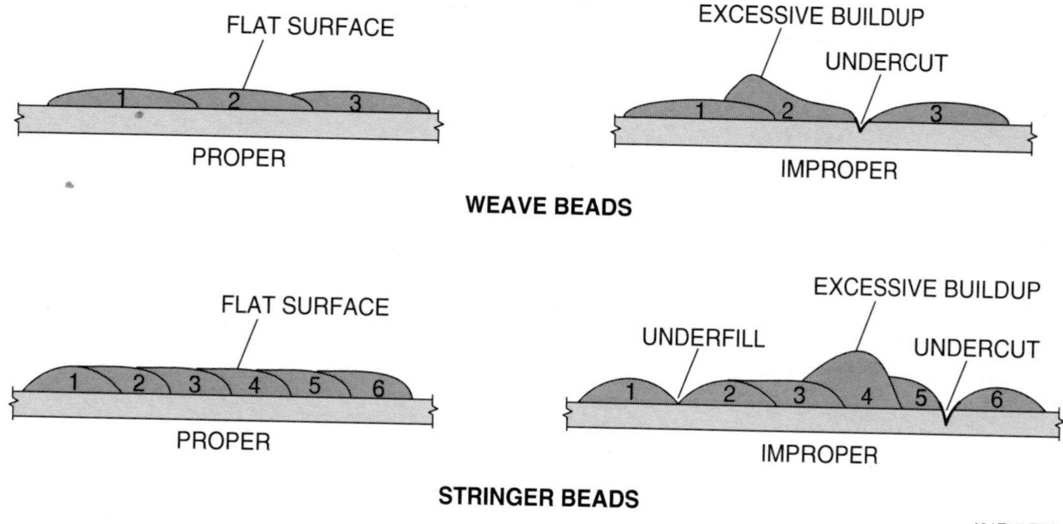

Figure 17 Proper and improper overlapping beads.

7.0.0 Fillet Welds

Fillet welds require little base metal preparation, except for the cleaning of the weld area and the removal of any excess material from cut surfaces. Any dross from cutting will cause porosity in the weld. For this reason, the codes require that this material is entirely removed prior to welding.

The most common fillet welds are made in lap and T-joints. The weld position for plate is determined by the axis of the weld and the orientation of the workpiece. The positions for fillet welding on plate (*Figure 18*) are flat (1F), horizontal (2F), vertical (3F), and overhead (4F). In the 1F and 2F positions, the weld axis can be inclined up to 15 degrees. Any weld axis inclination for the other positions varies with the rotational position of the weld face as specified in the AWS Standards.

Fillet welds can be concave or convex, depending on the WPS or site quality standards. Welding codes require a fillet weld to have a uniform concave or convex face, although a slightly nonuniform face is acceptable. The convexity of a fillet weld or individual surface bead must not exceed that permitted by the applicable code or standard. A fillet weld with profile defects is unacceptable and must be repaired (*Figure 19*).

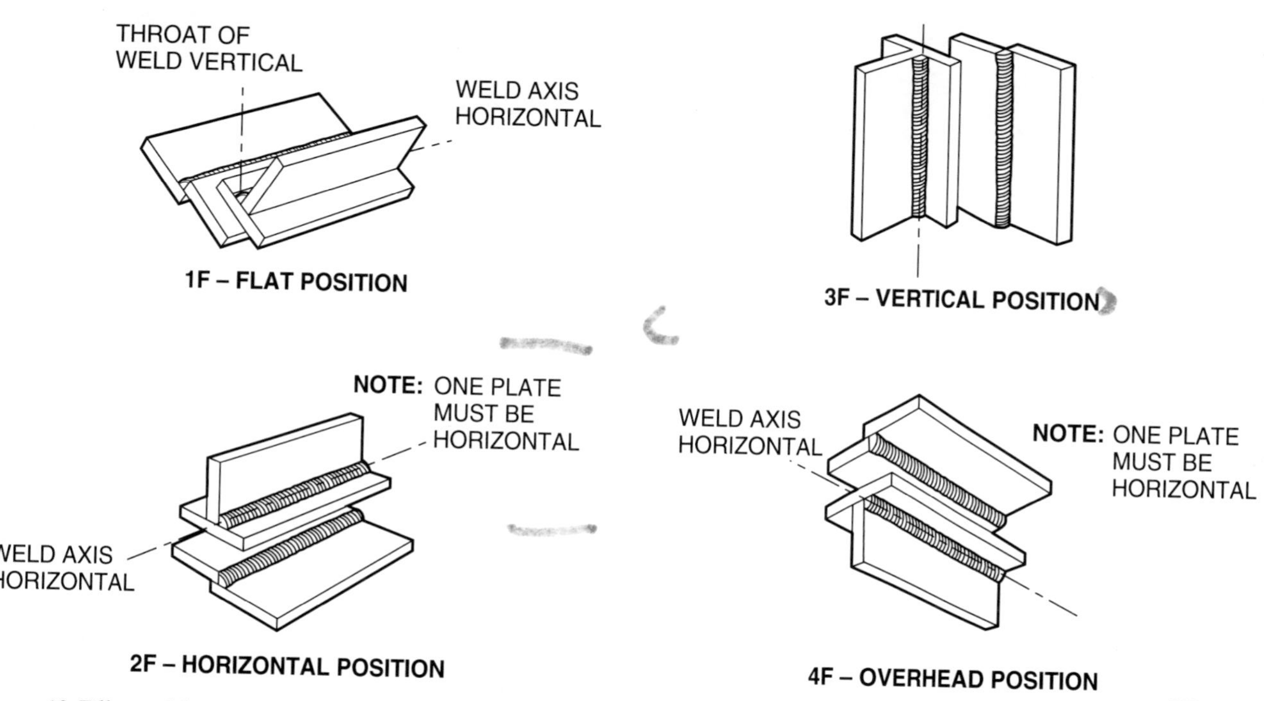

Figure 18 Fillet welding positions.

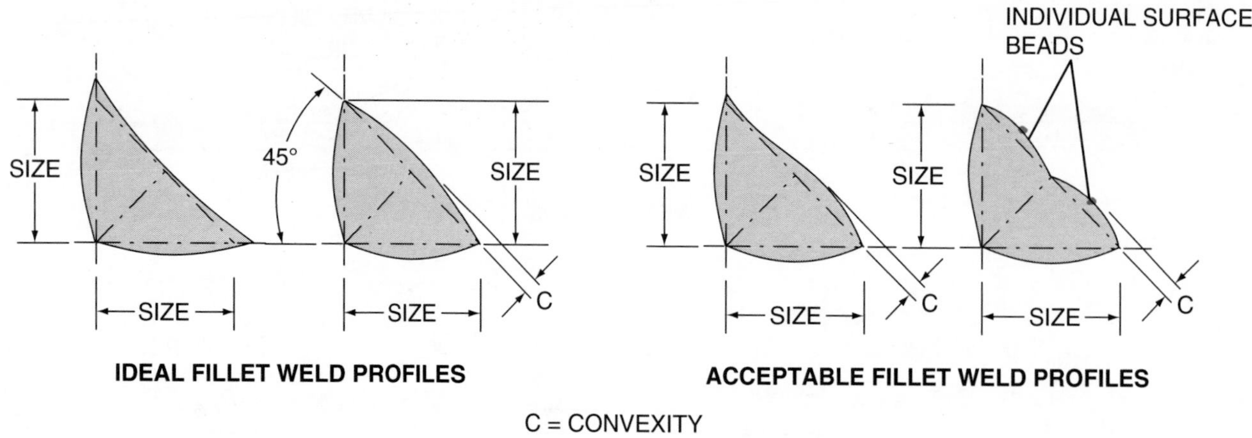

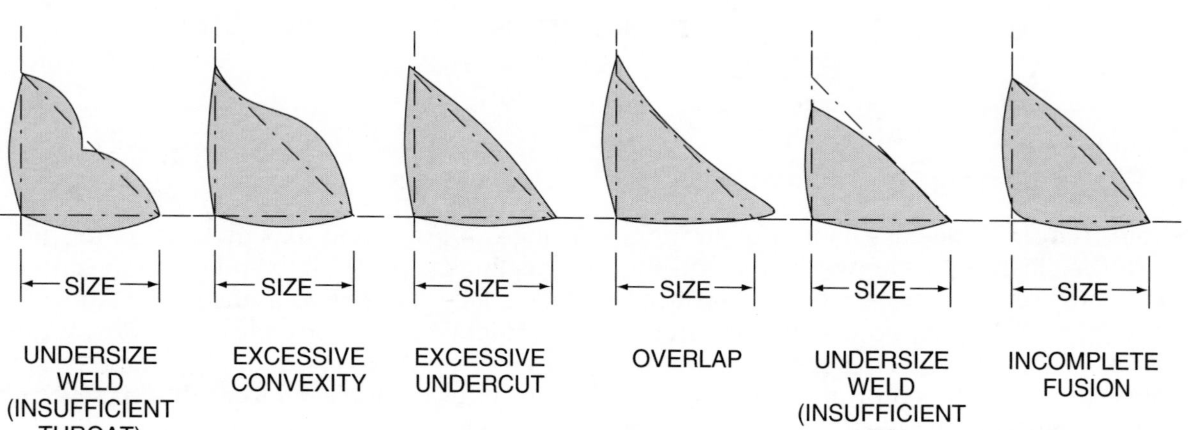

Figure 19 Acceptable and unacceptable fillet weld profiles.

7.1.0 Practicing Flat (1F) Position Fillet Welds

Practice 1F-position (flat) fillet welds by making multiple-pass (six-pass) convex fillet welds in a T-joint. Use an appropriate aluminum filler metal as directed by your instructor. When making flat fillet welds, pay close attention to the gun angle and travel speed. For the first bead, the gun work angle is vertical (45 degrees to both plate surfaces). Adjust the work angle for all subsequent beads (*Figure 20*). Clean each completed bead with a stainless steel wire brush before starting the next bead.

Follow these steps to make a flat fillet weld:

> **NOTE**
> In the following steps, clean all weld beads before beginning the next bead.

Step 1 Tack two plates together to form a T-joint for a fillet weld coupon (*Figure 21*). Clean the tack welds.

Step 2 Clamp or tack-weld the coupon in the 1F position (flat).

Step 3 Run the first bead along the root of the joint using a 45-degree work angle and a 5- to 10-degree push angle.

Step 4 Run the second bead along a toe of the first weld, overlapping about 75 percent of the first bead. Alter the work angle as shown in *Figure 20.* Use a 5- to 10-degree push angle with a slight oscillation.

Step 5 Run the third bead along the other toe of the first weld. Be sure to fill the groove created when the second bead was run. Use the work angle shown in *Figure 20* and a 5- to 10-degree push angle with a slight oscillation.

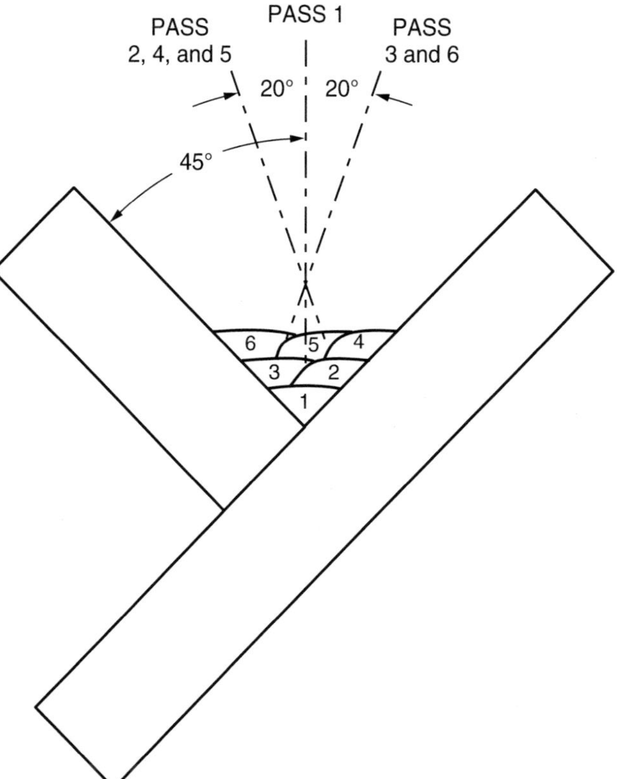

Figure 20 Multiple-pass 1F bead sequence and work angles.

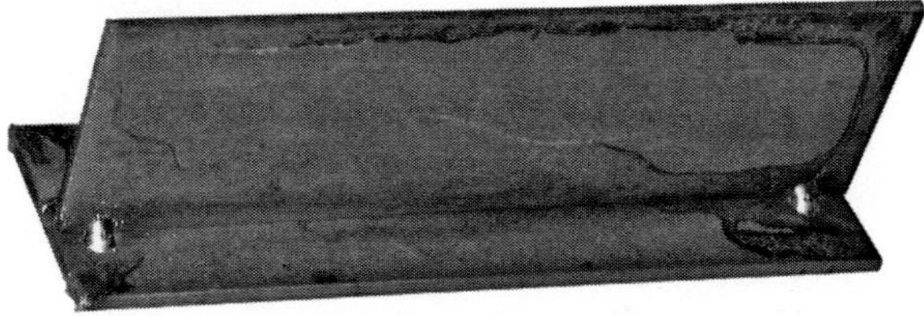

Figure 21 Fillet weld coupon.

Step 6 Run the fourth bead along the outside toe of the second weld, overlapping about half of the second bead. Use the work angle shown in *Figure 20* and a 5- to 10-degree push angle with a slight oscillation.

Step 7 Run the fifth bead along the inside toe of the fourth weld, overlapping about half of the fourth bead. Use the work angle shown in *Figure 20* and a 5- to 10-degree push angle with a slight oscillation.

Hot Tip

Tacking and Aligning Workpieces

When tacking workpieces together, position them to minimize distortion when the final welds are made. To achieve this, both sides of the workpieces are usually tacked with welds that are about ½" long. After the first tack weld, use a hammer or other tool to align the workpieces side-to-side and end-to-end. Then tack the opposite side. Tack the far ends of the workpieces in the same way. Intermediate tack welds can be made every 5" to 6" as needed to minimize lengthwise distortion.

Step 8 Run the sixth bead along the toe of the fifth weld. Be sure to fill the groove created when the fifth bead was run. Use the work angle shown in *Figure 20* and a 5- to 10-degree push angle with a slight oscillation.

Step 9 Have your instructor inspect the weld. The weld is acceptable if it has the following features:

- Uniform appearance on the bead face
- Craters and restarts filled to the full cross section of the weld
- Uniform weld size of ±1⁄16"
- Acceptable weld profile in accordance with the applicable code or standard
- Smooth transition with complete fusion at the toes of the weld
- No porosity
- No excessive undercut
- No overlap
- No inclusions
- No cracks

Hot Tip

T-Joint Heat Dissipation

In T-joints, the welding heat dissipates more rapidly in the thicker or non-butting member. On various bead passes, the arc may have to be slightly more concentrated on the thicker or nonbutting member to compensate for the heat loss.

7.2.0 Practicing Horizontal (2F) Position Fillet Welds

Practice horizontal (2F) fillet welding by placing multiple-pass fillet welds in a T-joint. Use an aluminum filler metal as directed by your instructor. When making horizontal fillet welds, pay close attention to the gun angles. For the first bead, the electrode work angle is 45 degrees. The work angle is adjusted for all other welds.

Follow these steps to make a horizontal fillet weld:

Step 1 Tack two plates together to form a T-joint for the fillet weld coupon. Clean the tack welds.

Step 2 Clamp or tack-weld the coupon in the horizontal position.

Step 3 Run the first bead along the root of the joint using a work angle of about 45 degrees and a 5- to 10-degree push angle (*Figure* 22).

Step 4 Clean the weld.

Step 5 Run the remaining passes at the proper work angles using a 5- to 10-degree push angle and a slight oscillation (*Figure* 22). Overlap each previous pass. Clean the weld after each pass.

Step 6 Have your instructor inspect the weld. The weld is acceptable if it has the following features:

- Uniform appearance on the bead face
- Craters and restarts filled to the full cross section of the weld
- Uniform weld size of ±1/16"
- Acceptable weld profile in accordance with the applicable code or standard
- Smooth transition with complete fusion at the toes of the weld
- No porosity
- No excessive undercut
- No overlap
- No inclusions
- No cracks

7.3.0 Practicing Vertical (3F) Position Fillet Welds

Practice vertical (3F) fillet welding by placing multiple-pass fillet welds in a T-joint. Use an aluminum filler metal as directed by your instructor.

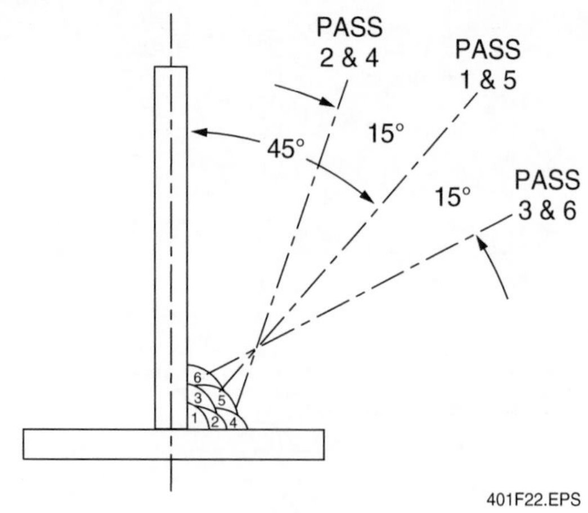

Figure 22 Multiple-pass 2F bead sequence and work angles.

Vertical welds are usually made by welding uphill from the bottom to the top using a gun push angle (up-angle). Because of the uphill welding and push angle, this type of welding is sometimes called vertical-up fillet welding. Either stringer or weave beads can be used for vertical welding. The site WPS or site quality standard will specify which technique to use on the job.

> **NOTE**
> Check with your instructor to see if you should practice stringer beads, weave beads, or both.

7.3.1 Weave Beads

Follow these steps to make an uphill fillet weld:

Step 1 Tack two plates together to form a T-joint for the fillet weld coupon.

Step 2 Clamp or tack-weld the coupon in the vertical position.

Step 3 Starting at the bottom, use a slightly oscillating motion to run the first bead along the root of the joint. Use a work angle of about 45 degrees and a 10- to 15-degree push angle. Pause in the weld puddle to fill the crater.

Step 4 Clean the weld.

Step 5 Run the remaining passes using a 10- to 15-degree push angle and a side-to-side weave technique with a 45-degree work angle (*Figure 23*). Use a slow motion across the face of the weld. Pause at each toe to penetrate and fill the crater. Clean the weld after each pass.

Step 6 Have your instructor inspect the weld. The weld is acceptable if it has the following features:

- Uniform appearance on the bead face
- Craters and restarts filled to the full cross section of the weld
- Uniform weld size of ±1⁄16"
- Acceptable weld profile in accordance with the applicable code or standard
- Smooth transition with complete fusion at the toes of the weld
- No porosity
- No excessive undercut
- No overlap
- No inclusions
- No cracks

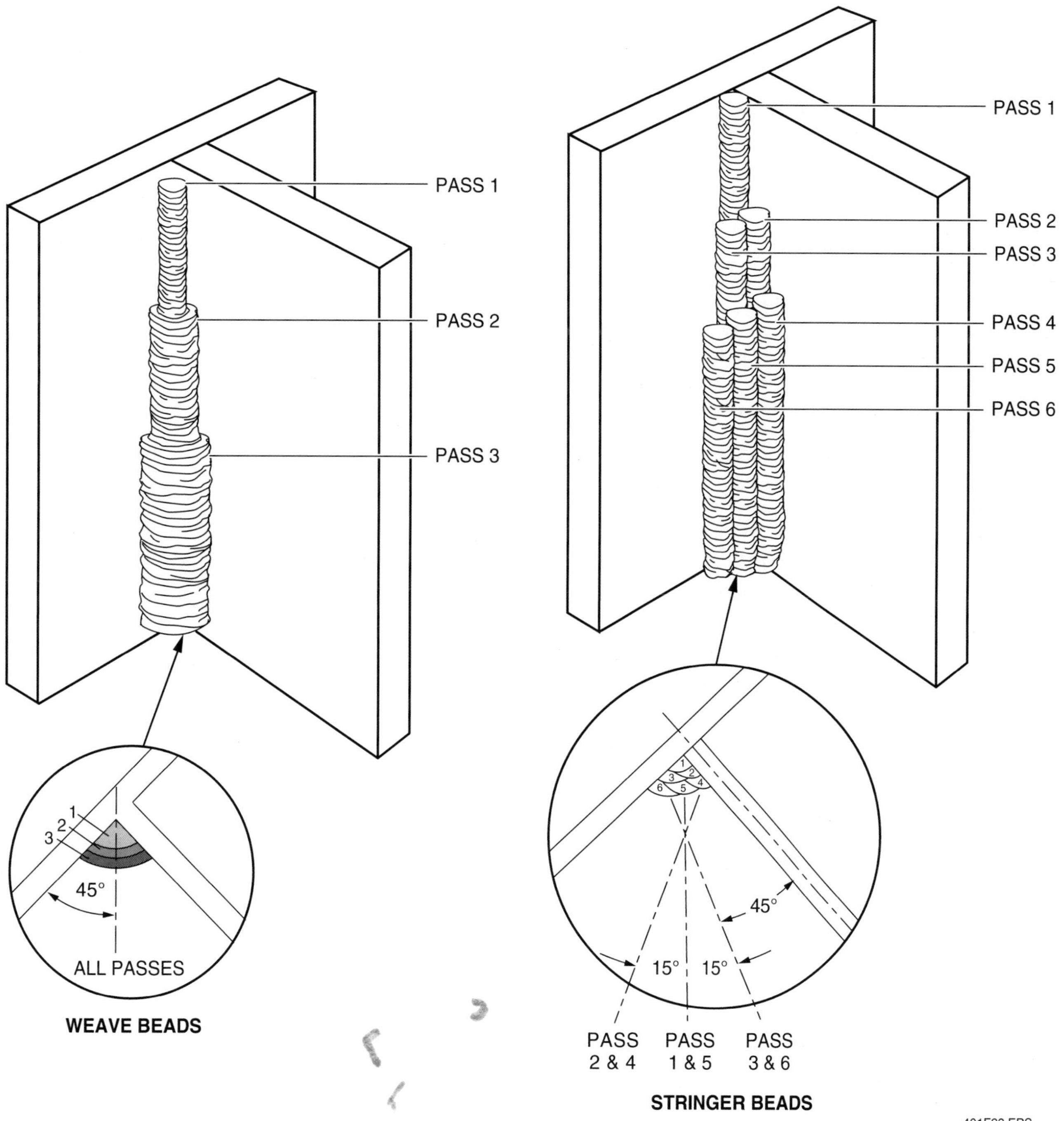

Figure 23 Multiple-pass 3F bead sequences and work angles for stringer and weave beads.

> **Hot Tip**
>
> ## Vertical Fillet Welds
>
> When making vertical fillet welds, pay close attention to the torch angles. The work angle is approximately 45 degrees for the first bead, and then it is adjusted for all other welds.

7.3.2 Stringer Beads

Repeat vertical fillet (3F) welding using stringer beads. Use a slightly oscillating motion, and pause slightly at each toe to prevent undercut. For stringer beads, use a 10- to 15-degree push angle and the required work angles (*Figure 23*).

7.4.0 Practicing Overhead (4F) Position Fillet Welds

Practice overhead (4F) fillet welding by welding multiple-pass fillet welds in a T-joint. Use an aluminum filler metal as directed by your instructor. When making overhead fillet welds, pay close attention to the gun angles. The work angle is about 45 degrees for the first bead, and then it is adjusted for all other welds.

Follow these steps to make an overhead fillet weld:

Step 1 Tack two plates together to form a T-joint for the fillet weld coupon.

Step 2 Clamp or tack-weld the coupon so it is in the overhead position.

Step 3 Run the first bead along the root of the joint. Use a work angle of about 45 degrees and a 5- to 10-degree push angle.

Step 4 Clean the weld.

Step 5 Run the remaining passes using a 5- to 10-degree push angle and the work angles shown in *Figure 24*. Overlap each previous pass. Clean the weld after each pass.

Step 6 Have your instructor inspect the weld. The weld is acceptable if it has the following features:

- Uniform appearance on the bead face
- Craters and restarts filled to the full cross section of the weld
- Uniform weld size of ±1⁄16"
- Acceptable weld profile in accordance with the applicable code or standard
- Smooth transition with complete fusion at the toes of the weld
- No porosity
- No excessive undercut
- No overlap
- No inclusions
- No cracks

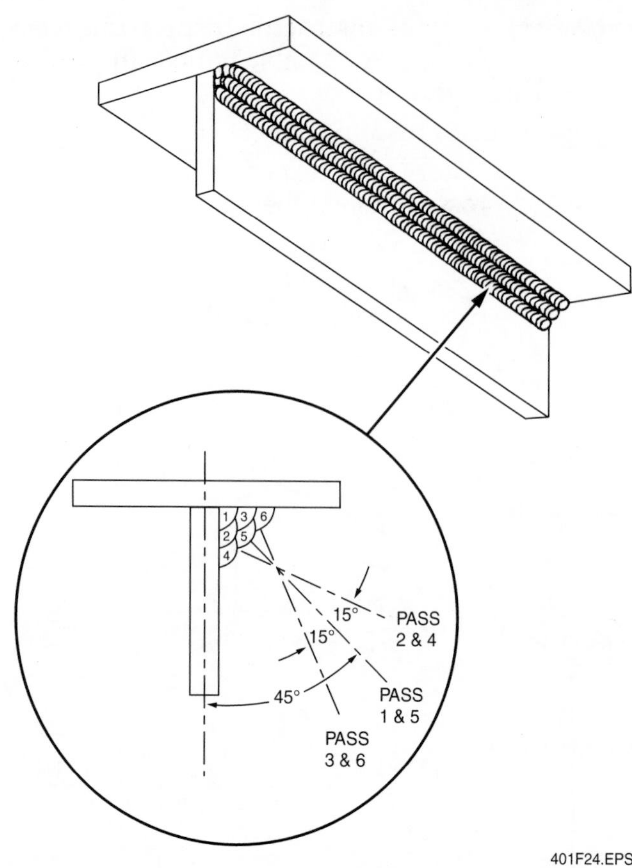

Figure 24 Multiple-pass 4F bead sequence and work angles.

8.0.0 V-Groove Plate Welds

The V-groove weld is a common groove weld made on plate and pipe. The backing method of welding an aluminum V-groove joint using GMAW is presented in this module. Practicing V-groove welds on plate is good preparation for making the more difficult pipe welds, which will be covered later.

V-groove welds with backing can be made in all positions. The weld position is determined by the axis of the weld and the orientation of the workpiece. Groove weld positions (*Figure 25*) are flat (1G), horizontal (2G), vertical (3G), and overhead (4G).

After completing the root pass, it should be cleaned and inspected. Use the proper cleaning method to remove the oxidation that tends to form along the joint. Inspect the root pass for the following features:

- Uniformly smooth face
- Complete fusion
- No excessive buildup
- No excessive undercut
- No porosity

Practice making GMAW V-groove welds with backing in the 1G, 2G, 3G, and 4G positions. Use the proper size of aluminum filler wire for the passes. Pay special attention to filling the crater at the termination of the weld. Clean each completed bead with a stainless steel wire brush before starting the next bead.

> **WARNING**
>
> Wear gloves when handling practice coupons, and be sure to use pliers when the coupons are hot.

V-groove welds with backing should be made with a slight reinforcement, not to exceed ⅛", and a gradual transition to the base metal at each toe. The root pass should have complete fusion to the backing. Groove welds must not have excessive reinforcement, any underfill, excessive undercut, or any overlap. If a groove weld has any of these defects (*Figure 26*), it must be repaired.

8.1.0 Practicing Flat (1G) Position V-Groove Welds

Follow these steps to practice making V-groove welds with backing in the flat (1G) position (*Figure 27*):

Step 1 Tack-weld the practice coupon together as explained earlier.

Step 2 Clamp or tack-weld the weld coupon in the flat position above the surface of the welding table.

Step 3 Use a 10- to 15-degree push angle and a 0-degree work angle to run the root pass with an appropriate aluminum filler metal.

Step 4 Clean the root pass.

Step 5 Run the remaining passes using a 5- to 10-degree push angle and the bead sequence and work angles shown in *Figure 27*. Clean the weld between each pass.

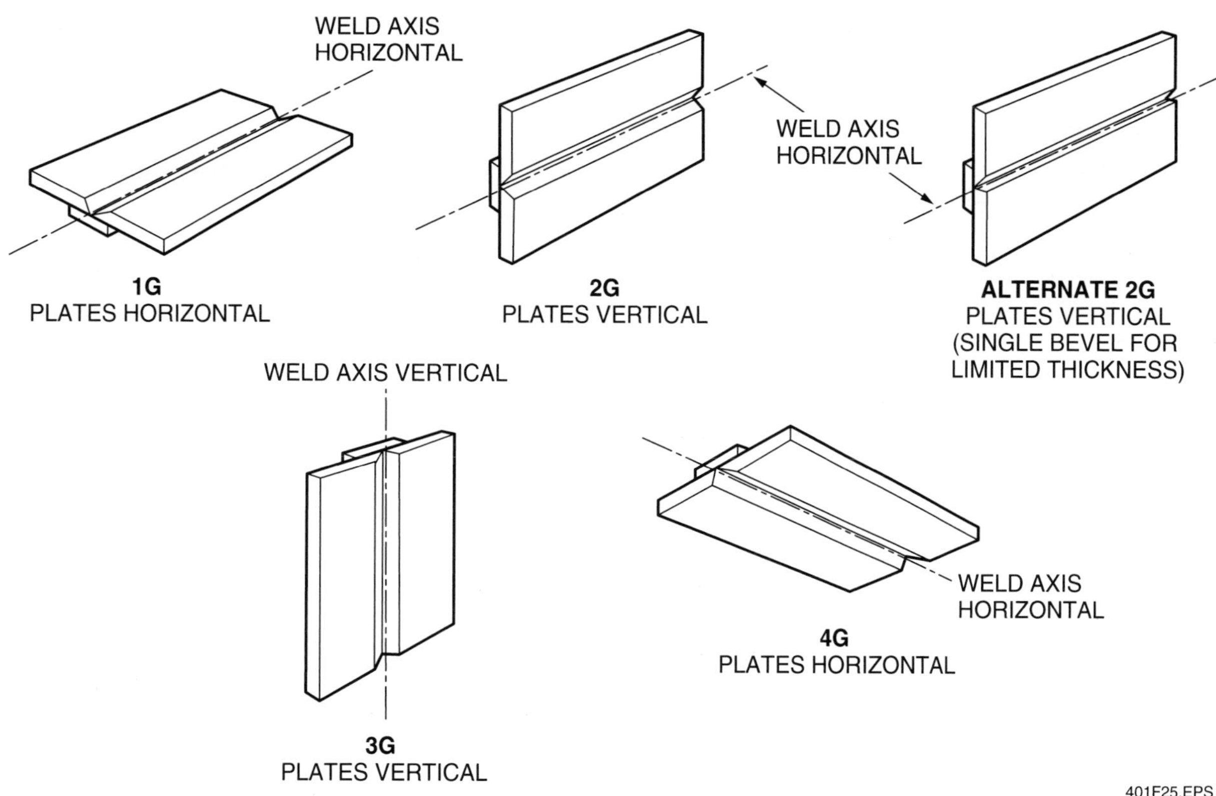

Figure 25 V-groove weld positions.

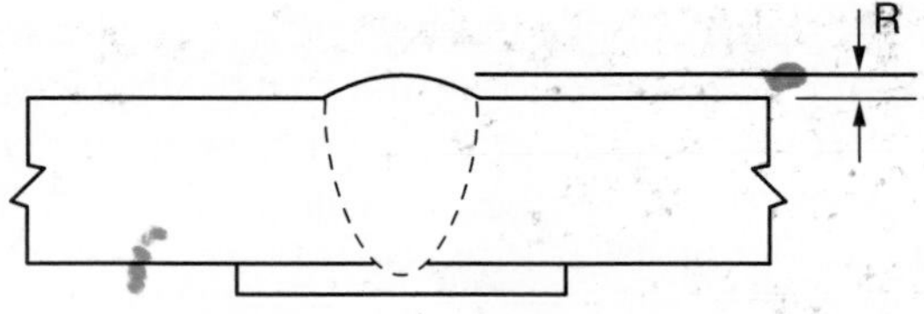

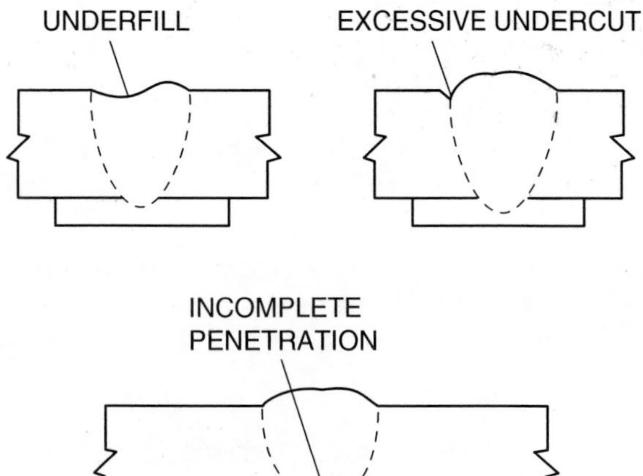

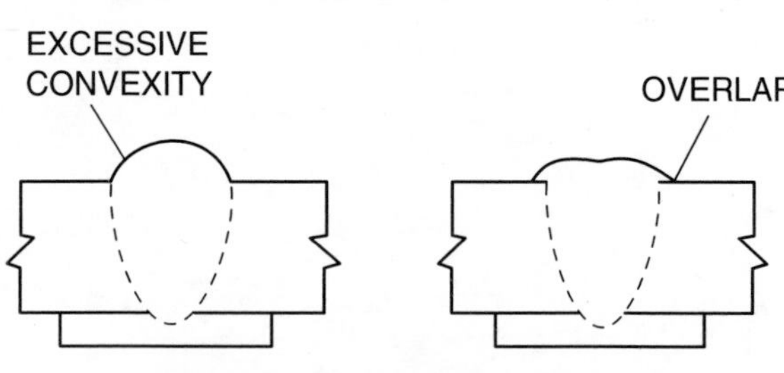

Figure 26 Acceptable and unacceptable V-groove weld profiles.

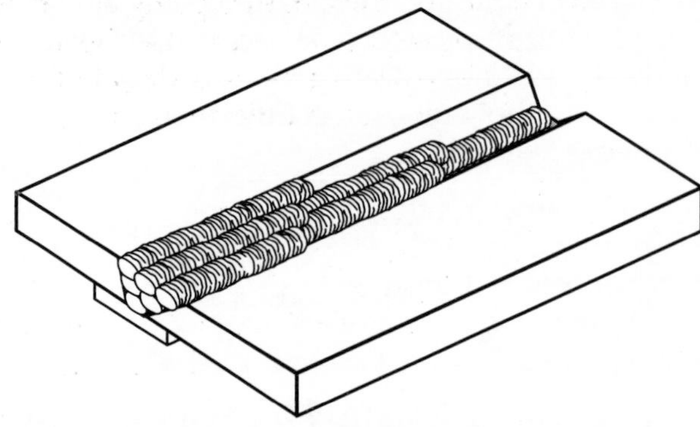

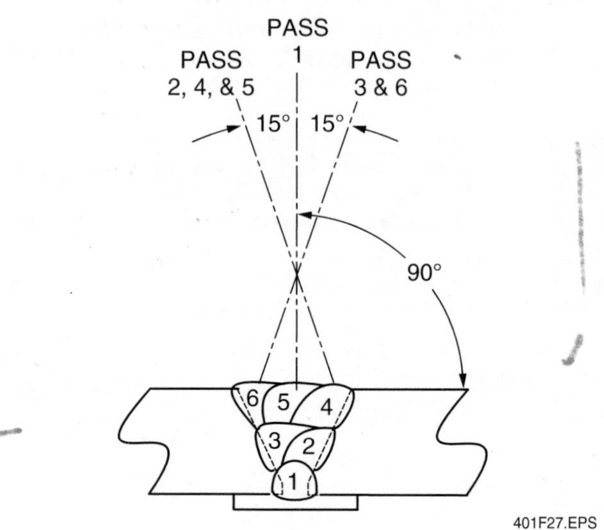

Figure 27 Multiple-pass 1G bead sequence and work angles.

Hot Tip

Specific V-Groove Requirements

Refer to your site's WPS for specific requirements on groove welds. The information in this module is provided only as a general guideline. The WPS or site quality standards must be followed for all welds. Check with your supervisor if you are unsure of the specifications for your application.

8.2.0 Practicing Horizontal (2G) Position V-Groove Welds

Follow these steps to practice making V-groove welds with backing in the horizontal (2G) position (*Figure 28*):

Step 1 Tack-weld the practice coupon together as explained earlier. Use the standard or alternate weld coupon as directed by your instructor.

Step 2 Clamp or tack-weld the coupon in the horizontal position.

Step 3 Use a 10- to 15-degree push angle to run the root pass with an appropriate aluminum filler metal. For Pass 1, use the work angle that corresponds to the alternate or standard joint shown in *Figure 28.*

Step 4 Clean the weld.

Step 5 Run the remaining passes using a 5- to 10-degree push angle and the bead sequences and work angles shown in *Figure 28*. Clean the weld between each pass.

8.3.0 Practicing Vertical (3G) Position V-Groove Welds

Follow these steps to practice making V-groove welds with backing in the vertical (3G) position (*Figure 29*):

> **NOTE**
>
> Stringer or weave beads may be used for vertical (3G) position welds, as specified by your instructor, WPS, or site quality standard.

Step 1 Tack-weld the practice coupon together as explained earlier.

Step 2 Clamp or tack-weld the coupon in the vertical position.

Step 3 Run the root pass uphill using an appropriate aluminum filler metal and a stringer bead. Use a push angle of 10 to 15 degrees and a 0-degree work angle for Pass 1 (*Figure 29)*.

Step 4 Clean the weld.

Step 5 Run the remaining passes uphill to complete the weld. Use a 5- to 10-degree push angle and the bead sequences and work angles shown in *Figure 29* for stringer or weave beads. Clean the weld between each pass.

Hot Tip

Cooling Practice Coupons

Cooling coupons with water can affect the mechanical properties of the base metal. Only practice coupons can be cooled with water. Never cool test coupons with water.

NOTE: THE ACTUAL NUMBER OF WELD BEADS WILL VARY DEPENDING ON THE PLATE THICKNESS.

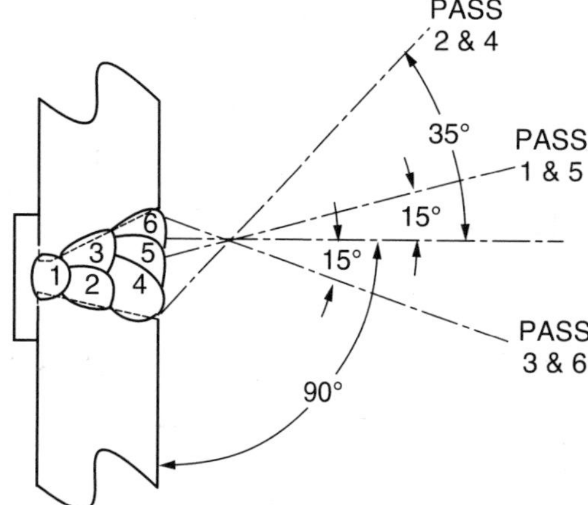

ALTERNATE JOINT REPRESENTATION

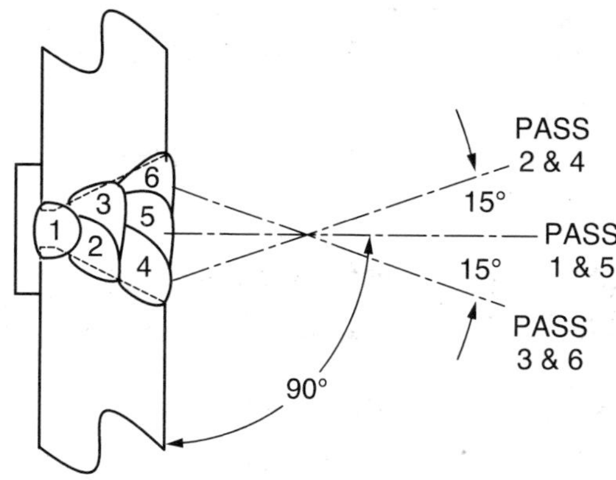

STANDARD JOINT REPRESENTATION

401F28.EPS

Figure 28 Multiple-pass 2G bead sequences and work angles.

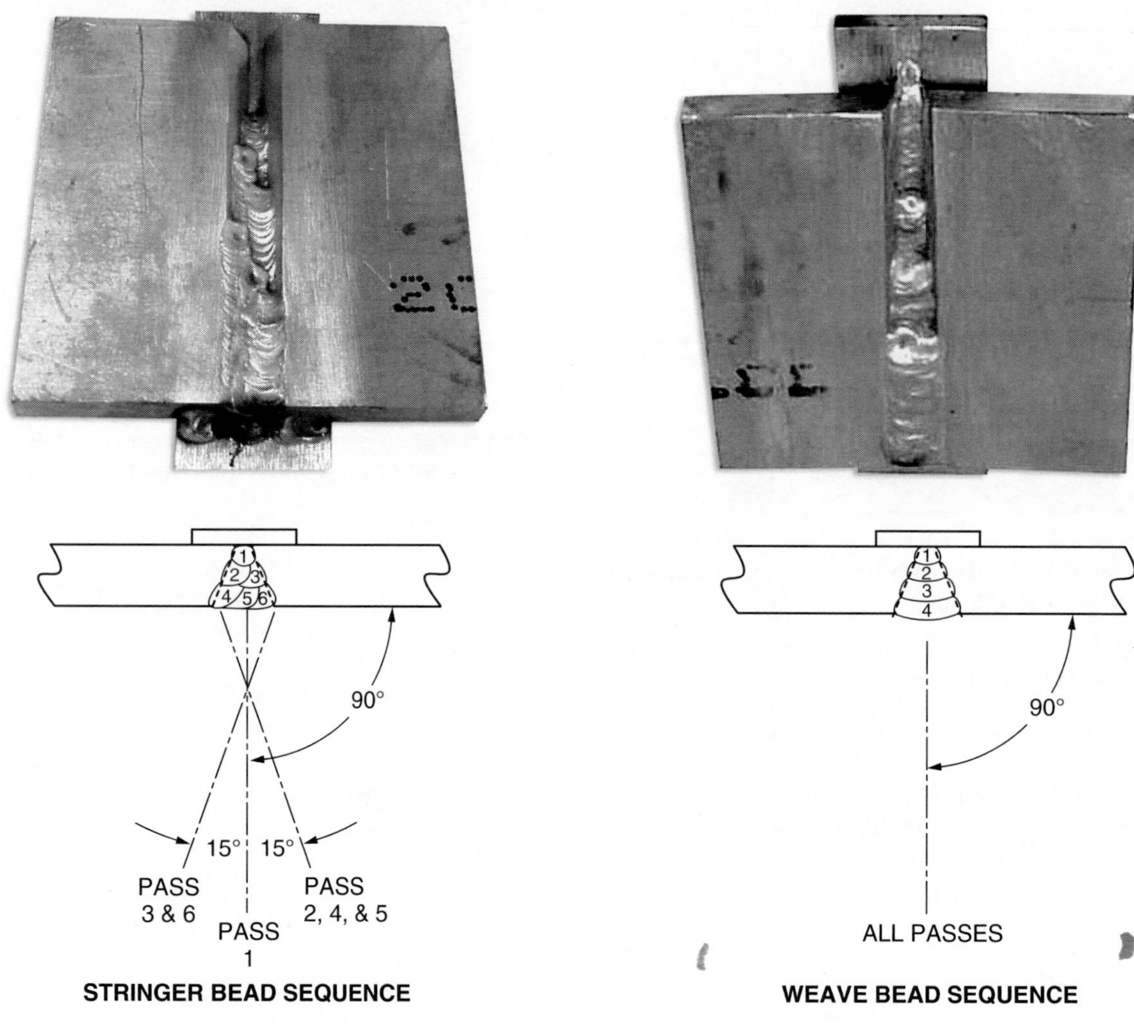

Figure 29 Multiple-pass 3G bead sequences and work angles.

8.4.0 Practicing Overhead (4G) Position V-Groove Welds

Follow these steps to practice making V-groove welds with backing in the overhead (4G) position (*Figure 30*):

Step 1 Tack-weld the practice coupon together as explained earlier.

Step 2 Clamp or tack-weld the coupon in the overhead position.

Step 3 Run the root pass using an appropriate aluminum filler metal. Use a 10- to 15-degree push angle and a 0-degree work angle for Pass 1 (*Figure 30).*

Step 4 Clean the weld.

Step 5 Run the remaining passes using a 5- to 10-degree push angle and the bead sequence and work angles shown in *Figure 30*. Clean the weld between each pass.

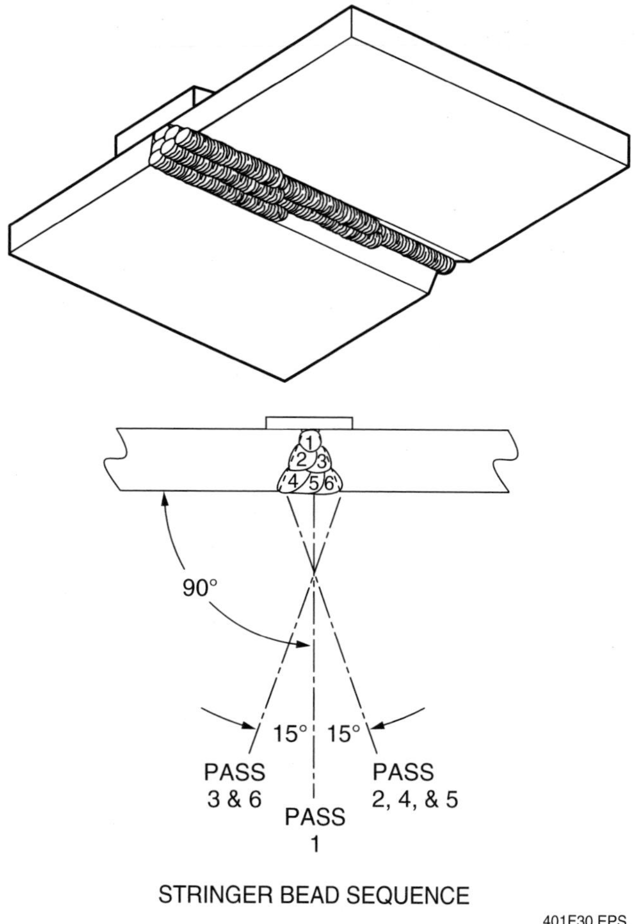

Figure 30 Multiple-pass 4G bead sequence and work angles.

SUMMARY

A few of the more difficult skills you must develop as a welder include setting up GMAW equipment, preparing the welding work area, and running stringer and weave beads. Another is making acceptable fillet and V-groove welds with backing on aluminum plate in all positions. Practice welding on aluminum plate until you can consistently produce acceptable welds as defined in the criteria for acceptance outlined in this module.

Review Questions

1. The GMAW process for aluminum uses which of the following for the filler metal?
   a. Short lengths of electrode
   b. Tin-coated aluminum rods
   c. A continuous wire electrode
   d. Nickel-coated aluminum rods

2. When welding aluminum, GTAW is frequently used for the root pass, while the process often used for the fill passes is _____.
   a. SMAW
   b. GMAW
   c. FCAW-G
   d. FCAW-S

3. Because of its strong inclination toward oxygen, aluminum oxidizes as soon as it is exposed to _____.
   a. water
   b. air
   c. carbon
   d. hydrogen

4. When welded, nonheat-treatable alloys may lose the effects of work hardening, which results in softening of the _____.
   a. heat-affected zone (HAZ)
   b. filler metal
   c. backing
   d. weld

5. Cast aluminum has a high silicon content for increased _____.
   a. oxidation resistance
   b. corrosion resistance
   c. fluidity
   d. strength

6. Special care should be taken to raise the temperature of cold aluminum slowly because rapid rises in temperature cause _____.
   a. distortion
   b. loss of strength
   c. loss of ductility
   d. condensation

7. When using preweld cleaning products, such as etchants or deoxidizers, the metal must be fully rinsed and completely dried to avoid creating a surface film of _____.
   a. hydrated oxide
   b. dehydrated oxide
   c. hydrated acetone
   d. aluminum dioxide

8. To remove oil and grease from aluminum coupons, use a(n) _____.
   a. approved solvent
   b. stainless steel buffer
   c. welding torch
   d. grinder

9. In GMAW, weld travel speed is measured in feet per minute.
   a. True
   b. False

10. Slower travel speeds result in deeper penetration and weld beads that are _____.
    a. higher
    b. shorter
    c. wider
    d. longer

11. In GMAW, the length of wire extending beyond the tip of the welding gun's contact tip is called the _____.
    a. setback
    b. stickout
    c. standoff distance
    d. electrode extension

12. In GMAW, the distance from the end of the gas nozzle to the end of the electrode is called the _____.
    a. setback
    b. stickout
    c. standoff distance
    d. electrode extension

Review Questions

13. If the weld spatter that accumulates on the gas nozzle is not removed, it will restrict the shielding gas flow and cause _____.
    a. porosity in the weld
    b. cracking in the weld
    c. cold fusion in the weld
    d. worm holes in the weld

14. Clean weld spatter from a gas nozzle with a _____.
    a. grinder
    b. round file
    c. chemical cleaner
    d. stainless steel brush

15. When making a weave bead, take care at the toes to _____.
    a. avoid pauses on the edges of the base metal
    b. touch the electrode to the base metal
    c. avoid oscillations on the base metal
    d. ensure proper fusion to the base metal

16. The junction where a new weld connects to and continues the bead of a previous weld is called a _____.
    a. juncture
    b. restart
    c. crater
    d. puddle

17. A weld termination normally leaves a _____.
    a. puddle
    b. crater
    c. keyhole
    d. padding

18. The overlapping of parallel beads to form a flat surface is called _____.
    a. fusing
    b. weaving
    c. transitioning
    d. padding

19. When making the first bead of a fillet weld in the overhead (4F) position, keep the gun work angle at _____.
    a. 15 degrees to the plate surface
    b. 45 degrees to the plate surface
    c. 55 degrees to the plate surface
    d. 90 degrees to the plate surface

20. When making V-groove welds with backing, the transition to the base metal at each toe should be _____.
    a. very noticeable
    b. intermittent
    c. gradual
    d. rapid

Trade Terms Introduced in This Module

Anodic coating: An artificial buildup of aluminum oxides on the surface of aluminum alloys that improves corrosion resistance. This coating is many times thicker than a coating of naturally occurring oxides.

Coalescence: The growing together or growth into one body of the materials being joined.

Ingot: A mass of metal shaped for convenient storage and transport.

Liquation: The process of separation by melting, such as in an alloy when one element melts while the others remain solid.

Additional Resources

This module is intended to present thorough resources for task training. The following references are suggested for further study. These are optional materials for continued education rather than for task training.

AWS D1.2/D1.2M:2008 Structural Welding Code – Aluminum. Miami, FL: American Welding Society.

AWS PRGQA The Practical Reference Guide for High Quality Fusion Welding of Aluminum. Miami, FL: American Welding Society, 2001.

AWS PRGWA The Practical Reference Guide to Welding Aluminum in Commercial Applications. Miami, FL: American Welding Society, 2002.

Lincoln Electric website: http://www.lincolnelectric.com offers sources for products and training.

The Procedure Handbook of Arc Welding. Cleveland, OH: The James F. Lincoln Arc Welding Foundation, 2000.

Welding Aluminum: Theory and Practice. New York, NY: The Aluminum Association, 2002.

Figure Credits

Lincoln Electric Company, Module opener, 401F02

Tony Anderson, 401F03

Terry Lowe, 401F07 (photo), 401F14, 401F15 (photo), 401F28 (photo), 401F29 (photos)

Topaz Publications, Inc., 401F21

PERFORMANCE ACCREDITATION TASKS

The Performance Accreditation Tasks (PATs) correspond to and support learning objectives in the *AWS EG2.0:2006 Curriculum Guide for the Training of Welding Personnel; Level I – Entry Welder*.

PATs provide specific acceptable criteria for performance and help to ensure a true competency-based welding program for students.

The following tasks are designed to evaluate your ability to run stringer beads, weave beads, overlapping beads, and to make fillet and V-groove welds with GMAW equipment in four standard test positions using aluminum filler wire of the appropriate diameter and shielding gas. Perform each task when you are instructed to do so by your instructor. As you complete each task, show it to your instructor for evaluation. Do not proceed to the next task until told to do so by your instructor. For AWS 2G and 5G certifications, refer to *AWS EG3.0:1996 Guide for the Training and Qualification of Welding Personnel; Level II – Advanced Welder* for bend test requirements. For AWS 6G certifications, refer to *AWS EG4.0:1996 Guide for the Training and Qualification of Welding Personnel; Level III – Expert Welder* for bend test requirements.

WELD A PAD ON ALUMINUM PLATE IN THE FLAT (1G) POSITION USING GMAW STRINGER BEADS

As directed by the instructor, use the GMAW process with the appropriate aluminum filler wire to make the following welds on aluminum plate: stringer beads, weave beads, weld restarts, weld terminations, and overlapping beads.

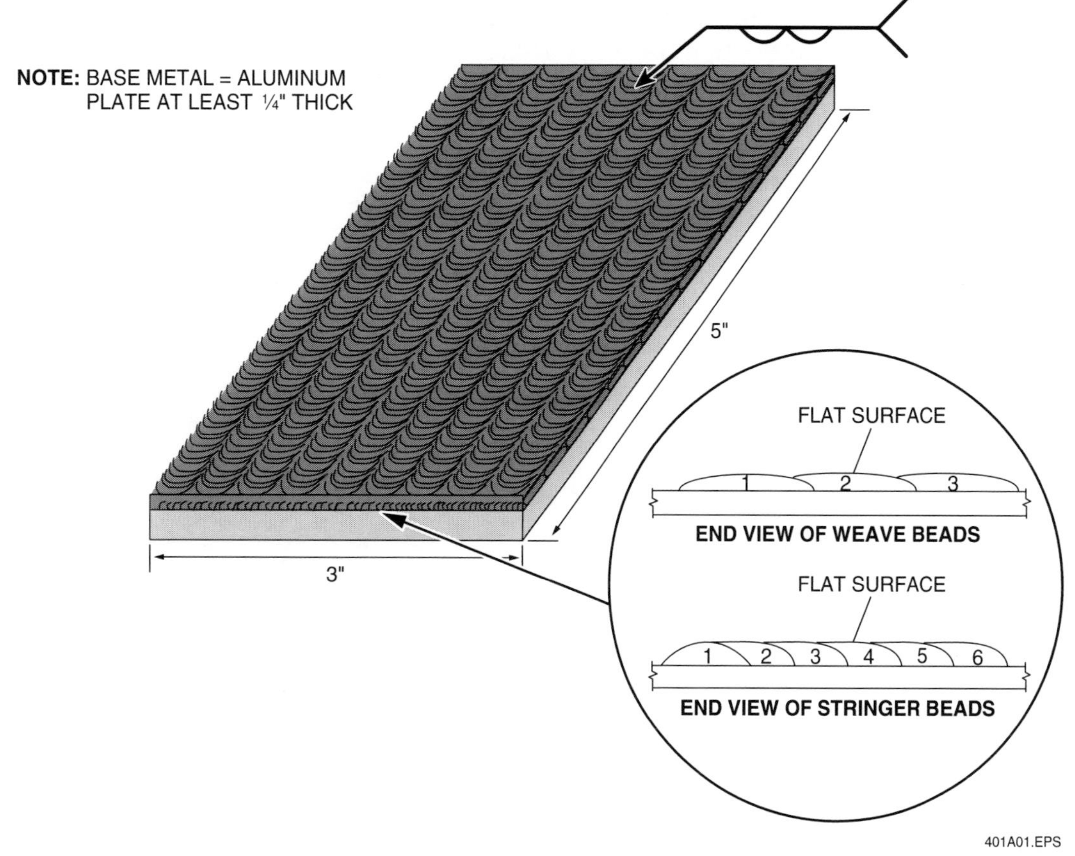

Criteria for Acceptance

- Uniform rippled appearance on the bead face ______
- Craters and restarts filled to the full cross section of the weld ______
- Uniform weld width ±1/16" ______
- Acceptable weld profile in accordance with the acceptable code or standard ______
- Smooth flat transition with complete fusion at the toes of the weld ______
- No porosity ______
- No excessive undercut ______
- No inclusions ______
- No cracks ______

MAKE MULTIPLE-PASS FILLET WELDS ON ALUMINUM PLATE IN THE FLAT (1F) POSITION

As directed by the instructor, use the GMAW process with the appropriate aluminum wire to make a six-pass fillet weld using stringer beads on aluminum plate, as shown.

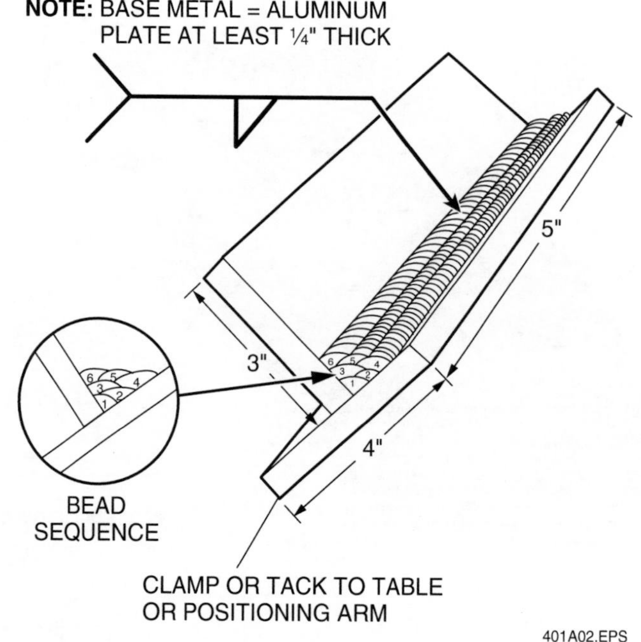

Criteria for Acceptance

- Uniform rippled appearance on the bead face ____________
- Craters and restarts filled to the full cross section of the weld ____________
- Uniform weld size ±1/16" ____________
- Smooth flat transition with complete fusion at the toes of the welds ____________
- Acceptable weld profile in accordance with the applicable code or standard ____________
- No porosity ____________
- No excessive undercut ____________
- No overlap ____________
- No inclusions ____________

MAKE MULTIPLE-PASS FILLET WELDS ON ALUMINUM PLATE IN THE HORIZONTAL (2F) POSITION

As directed by the instructor, use the GMAW process with the appropriate aluminum wire to make a six-pass fillet weld using stringer beads on aluminum plate, as shown.

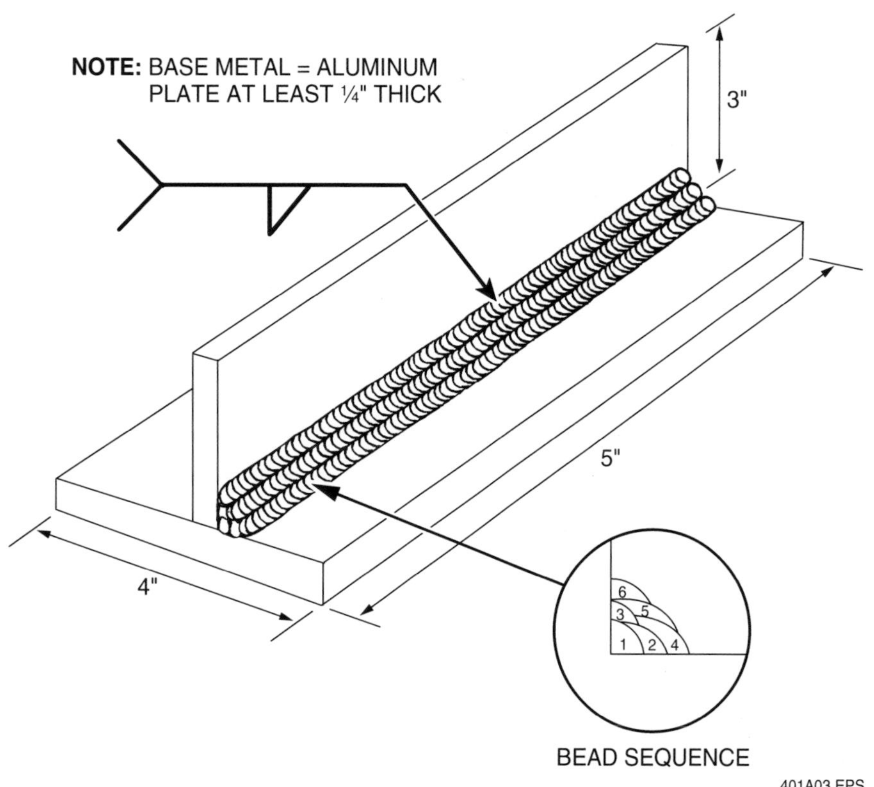

Criteria for Acceptance

- Uniform rippled appearance on the bead face ____________
- Craters and restarts filled to the full cross section of the weld ____________
- Uniform weld size ±1⁄16" ____________
- Smooth flat transition with complete fusion at the toes of the welds ____________
- Acceptable weld profile in accordance with the applicable code or standard ____________
- No porosity ____________
- No excessive undercut ____________
- No overlap ____________
- No inclusions ____________

MAKE MULTIPLE-PASS FILLET WELDS ON ALUMINUM PLATE IN THE VERTICAL (3F) POSITION

As directed by the instructor, use the GMAW process with the appropriate aluminum wire to make a six-pass fillet weld using stringer beads on aluminum plate, as shown.

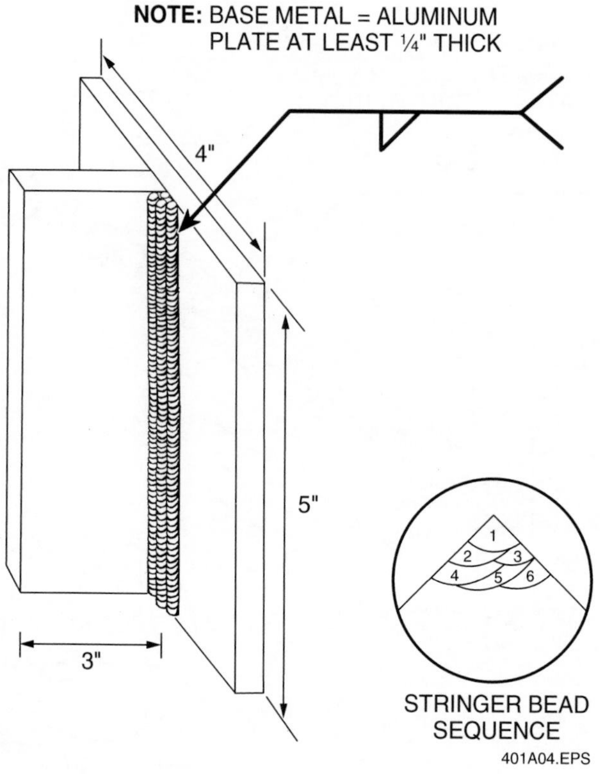

Criteria for Acceptance

- Uniform rippled appearance on the bead face ____________
- Craters and restarts filled to the full cross section of the weld ____________
- Uniform weld width ±1⁄16" ____________
- Acceptable weld profile in accordance with the applicable code or standard ____________
- Smooth flat transition with complete fusion at the toes of the weld ____________
- No porosity ____________
- No excessive undercut ____________
- No inclusions ____________
- No cracks ____________

MAKE MULTIPLE-PASS FILLET WELDS ON ALUMINUM PLATE IN THE OVERHEAD (4F) POSITION

As directed by the instructor, use the GMAW process with the appropriate aluminum wire to make a six-pass fillet weld using stringer beads on aluminum plate, as shown.

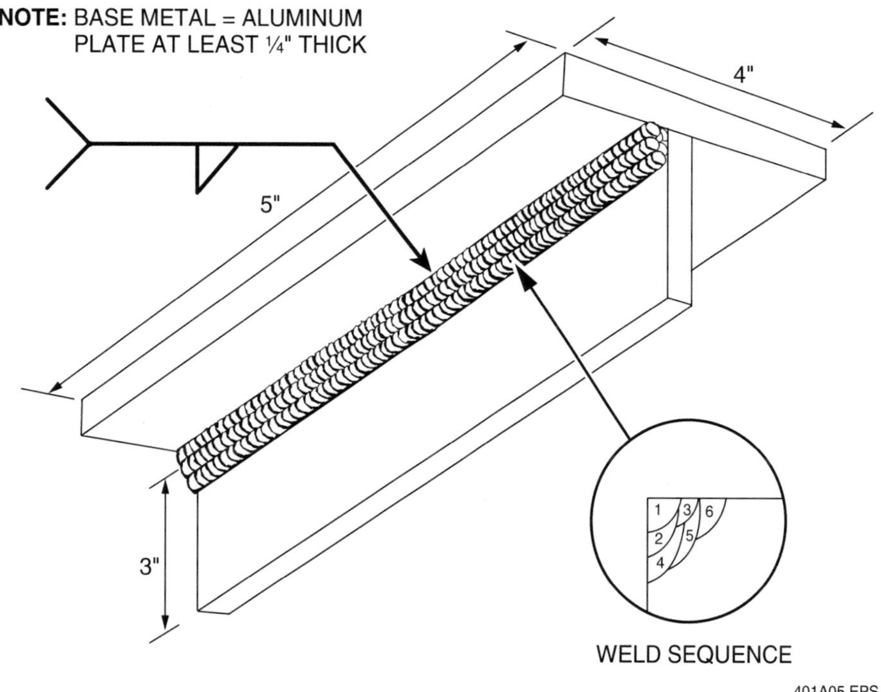

Criteria for Acceptance

- Uniform rippled appearance on the bead face ________
- Craters and restarts filled to the full cross section of the weld ________
- Uniform weld width ±$\frac{1}{16}$" ________
- Acceptable weld profile in accordance with the applicable code or standard ________
- Smooth flat transition with complete fusion at the toes of the weld ________
- No porosity ________
- No excessive undercut ________
- No inclusions ________
- No cracks ________

MAKE MULTIPLE-PASS V-GROOVE WELDS ON ALUMINUM PLATE WITH BACKING IN THE FLAT (1G) POSITION

As directed by the instructor, use the GMAW process with the appropriate aluminum wire to make a multiple-pass groove weld using stringer beads on aluminum plate with backing, as shown.

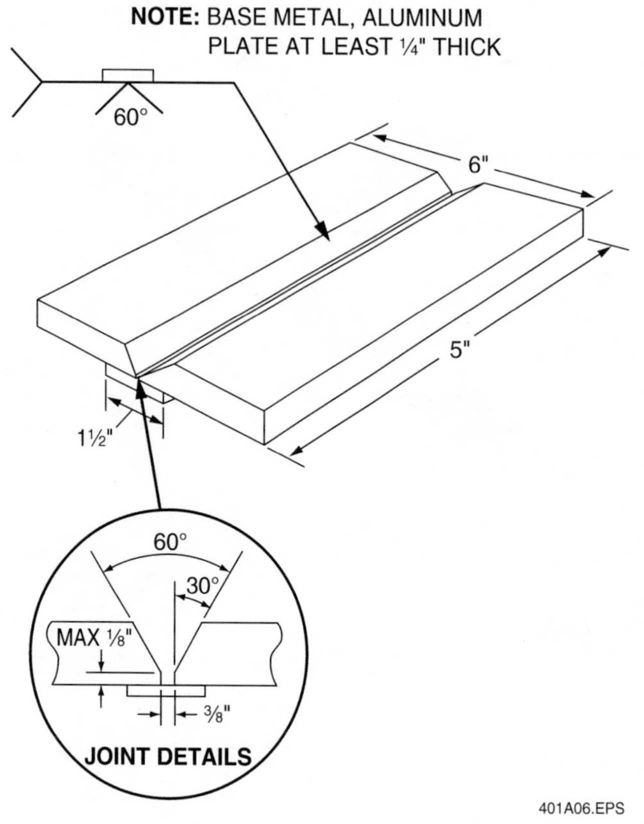

401A06.EPS

Criteria for Acceptance

- Uniform rippled appearance on the bead face ________
- Craters and restarts filled to the full cross section of the weld ________
- Acceptable weld profile in accordance with the applicable code or standard ________
- Smooth flat transition with complete fusion at the toes of the weld ________
- No porosity ________
- No excessive undercut ________
- No inclusions ________
- No cracks ________

MAKE MULTIPLE-PASS V-GROOVE WELDS ON ALUMINUM PLATE WITH BACKING IN THE HORIZONTAL (2G) POSITION

As directed by the instructor, use the GMAW process with the appropriate aluminum wire to make a multiple-pass groove weld on aluminum plate with backing, as shown.

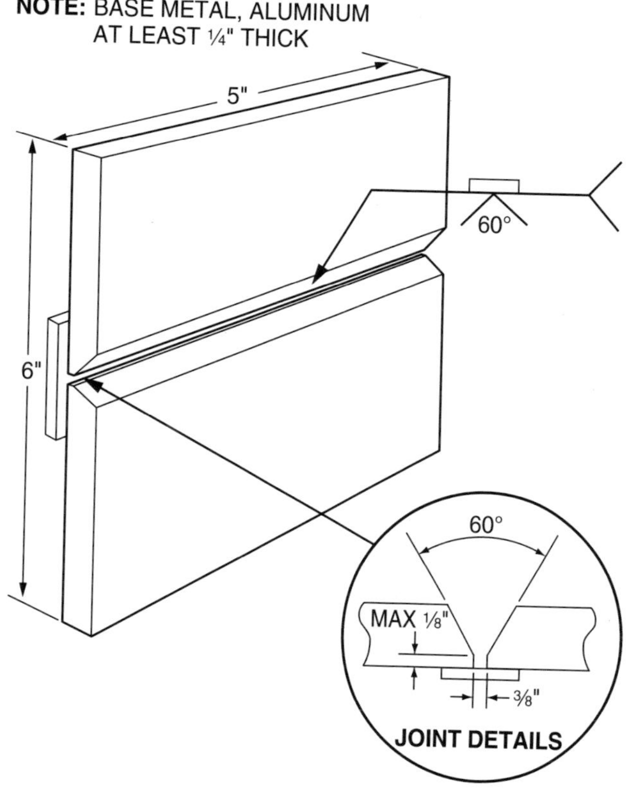

Criteria for Acceptance

- Uniform rippled appearance on the bead face
- Craters and restarts filled to the full cross section of the weld ______
- Acceptable weld profile in accordance with the applicable code or standard ______
- Smooth flat transition with complete fusion at the toes of the weld ______
- No porosity ______
- No excessive undercut ______
- No inclusions ______
- No cracks ______

______

MAKE MULTIPLE-PASS V-GROOVE WELDS ON ALUMINUM PLATE WITH BACKING IN THE VERTICAL (3G) POSITION

As directed by the instructor, use the GMAW process with the appropriate aluminum wire to make a multiple-pass groove weld on aluminum plate with backing, as shown.

Note: Run the root vertical up.

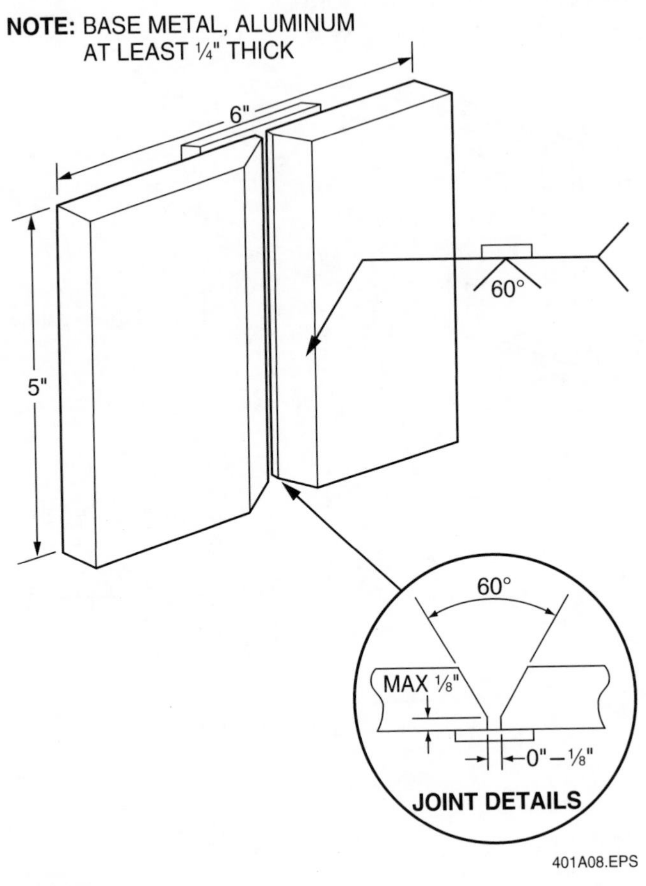

Criteria for Acceptance

- Uniform rippled appearance on the bead face ______
- Craters and restarts filled to the full cross section of the weld ______
- Acceptable weld profile in accordance with the applicable code or standard ______
- Smooth flat transition with complete fusion at the toes of the weld ______
- No porosity ______
- No excessive undercut ______
- No inclusions ______
- No cracks ______

MAKE MULTIPLE-PASS V-GROOVE WELDS ON ALUMINUM PLATE WITH BACKING IN THE OVERHEAD (4G) POSITION

As directed by the instructor, use the GMAW process with the appropriate aluminum wire to make a multiple-pass groove weld on aluminum plate with backing, as shown.

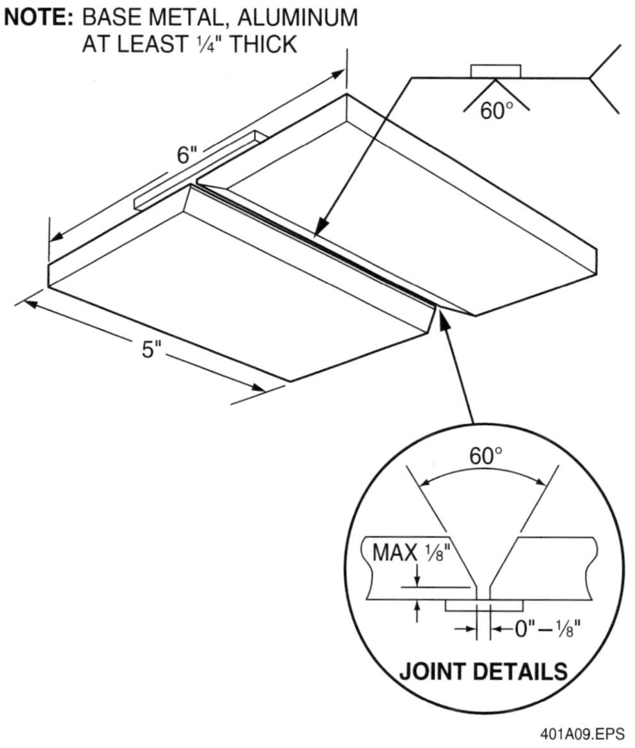

Criteria for Acceptance

- Uniform rippled appearance on the bead face ____________
- Craters and restarts filled to the full cross section of the weld ____________
- Acceptable weld profile in accordance with the applicable code or standard ____________
- Smooth flat transition with complete fusion at the toes of the weld ____________
- No porosity ____________
- No excessive undercut ____________
- No inclusions ____________
- No cracks ____________

NCCER CURRICULA — USER UPDATE

NCCER makes every effort to keep its textbooks up-to-date and free of technical errors. We appreciate your help in this process. If you find an error, a typographical mistake, or an inaccuracy in NCCER's curricula, please fill out this form (or a photocopy), or complete the online form at **www.nccer.org/olf**. Be sure to include the exact module ID number, page number, a detailed description, and your recommended correction. Your input will be brought to the attention of the Authoring Team. Thank you for your assistance.

Instructors – If you have an idea for improving this textbook, or have found that additional materials were necessary to teach this module effectively, please let us know so that we may present your suggestions to the Authoring Team.

NCCER Product Development and Revision

13614 Progress Blvd., Alachua, FL 32615

Email: curriculum@nccer.org

Online: www.nccer.org/olf

❑ Trainee Guide ❑ AIG ❑ Exam ❑ PowerPoints Other ______________________________

Craft / Level: Copyright Date:

Module ID Number / Title:

Section Number(s):

Description:

Recommended Correction:

Your Name:

Address:

Email: Phone:

GTAW – Aluminum Plate

29402-10

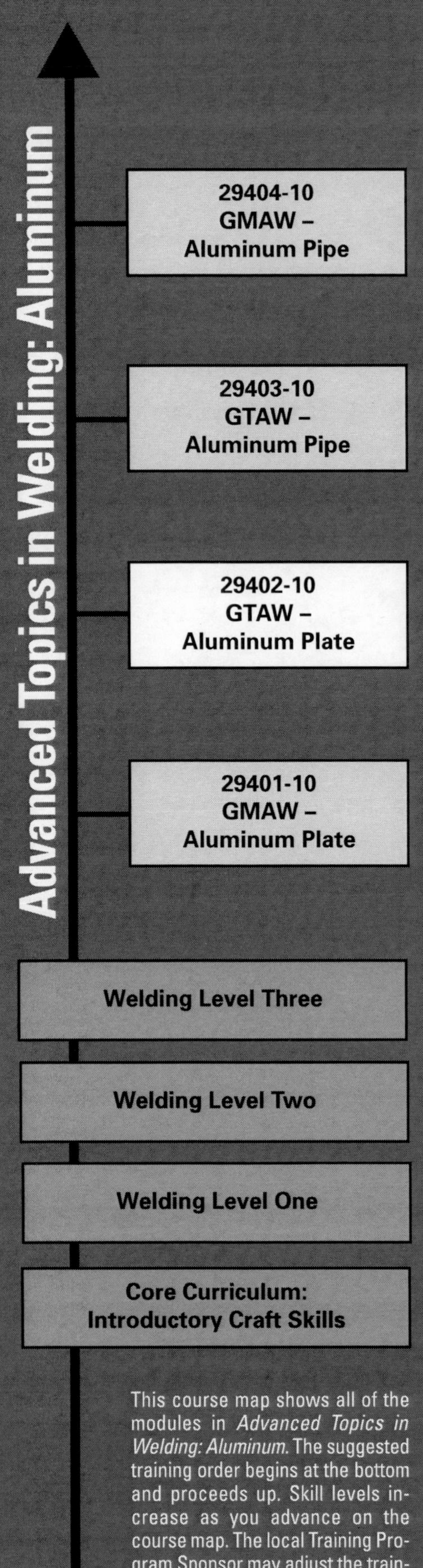

This course map shows all of the modules in *Advanced Topics in Welding: Aluminum.* The suggested training order begins at the bottom and proceeds up. Skill levels increase as you advance on the course map. The local Training Program Sponsor may adjust the training order.

Objectives

When you have completed this module, you will be able to do the following:

1. Explain GTAW and how to set up GTAW equipment to weld aluminum plate.
2. Explain and demonstrate GTAW techniques used to weld aluminum.
3. Build a pad with stringer beads and weave beads, using GTAW equipment and aluminum filler metal.
4. Make multiple-pass fillet welds on aluminum plate in the following positions, using GTAW equipment and aluminum filter metal:
   - 1F
   - 2F
   - 3F
   - 4F
5. Make multiple-pass V-groove welds on aluminum plate with backing in the following positions, using GTAW equipment and aluminum filler metal:
   - 1G
   - 2G
   - 3G
   - 4G

Trade Term

Helium/argon mixture

Prerequisites

Before you begin this module, it is recommended that you successfully complete *Core Curriculum*; *Welding Level One*; *Welding Level Two*; *Welding Level Three*; and *Advanced Topics in Welding: Aluminum*, Module 29401-10.

Contents

Topics to be presented in this module include:

Figures and Tables

1.0.0 INTRODUCTION

Gas tungsten arc welding (GTAW) is an arc welding process used for making high-quality welds on aluminum. In the past, this method of welding was called tungsten inert gas (TIG) welding. Tungsten refers to the tungsten electrode that delivers the arc. *Figure 1* shows the GTAW process and equipment.

This module covers GTAW equipment setup. It also explains how to prepare aluminum plate for fillet welds in the 1F, 2F, 3F, and 4F positions and V-groove welds in the 1G, 2G, 3G, and 4G positions.

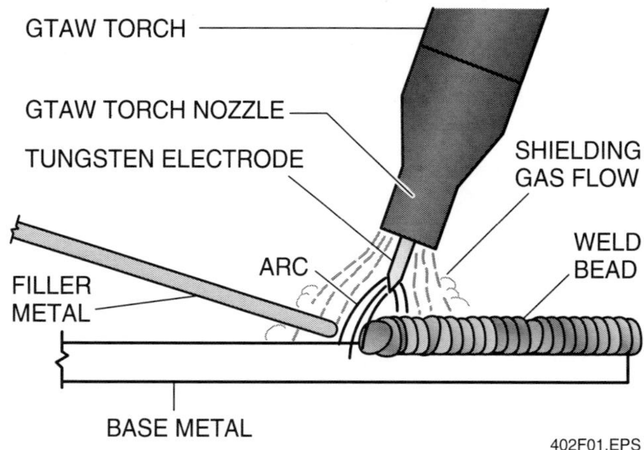

Figure 1 GTAW process.

1.1.0 Safety Practices

The following is a summary of safety procedures and practices that must be observed when welding. Keep in mind that this is only a summary; complete safety coverage is provided in the Level One module, *Welding Safety*. If you have not completed that module, do so before continuing. Above all, always use the proper protective clothing and equipment when welding.

1.1.1 Protective Clothing and Equipment

Welding work creates flying debris, such as sparks or small chunks of hot metal. Anyone welding or assisting a welder must use the proper protective clothing and equipment. The following list provides protective clothing and equipment guidelines:

- Always use safety glasses with a full face shield or a helmet. The glasses, face shield, or helmet lens must have the proper light-reducing shade for the type of welding being performed. Never directly or indirectly view an electric arc without using a properly tinted lens.
- Wear proper protective leather and/or flame retardant clothing and welding gloves. They will protect you from flying sparks, molten metal, and heat.

> **Hot Tip**
>
> **Dimensions and Specifications**
>
> The dimensions and specifications in this module are designed to be representative of codes in general and are not specific to any certain code. Always follow the proper codes for your site.

- Wear 8-inch or taller high-top safety shoes or boots. Make sure that the tongue and lace area is covered by a pant leg. Sometimes the tongue and lace area is exposed or the footwear must be protected from burn marks. In those cases, wear leather spats under the pants or chaps and over the front and top of the footwear.
- Wear a solid material (non-mesh) hat with a bill pointing to the back or toward the ear closest to the welding. This will give added protection. If much overhead welding is required, use a full leather hood with a welding faceplate and the correct tinted lens. If a hard hat is required, use one that allows the attachment of both rear deflector material and a face shield.
- Wear earmuffs or earplugs to protect ear canals from sparks.

1.1.2 Fire/Explosion Prevention

Welding work includes the cutting, grinding, and welding of metal. All of these actions generate heat and often produce flying sparks. The heat and flying sparks can be the cause of fires and explosions. Welders must use extreme care to protect both themselves and others near their work. The following list provides fire and explosion protection guidelines:

- Never carry matches or gas-filled lighters in your pockets. Sparks can cause the matches to ignite or the lighter to explode, causing serious injury.
- Never perform any type of heating, cutting, or welding until a hot work permit has been obtained and an approved fire watch established. Most work-site fires in these types of operations are started by cutting torches.

- Never use oxygen to blow dust or dirt from clothing. The oxygen can remain trapped in the fabric for a time. If a spark hits the clothing during this period, the clothing can burn rapidly and violently.
- Make sure that any flammable material in the work area is either moved to a safe area or shielded by a fire-resistant covering. Approved fire extinguishers must be available before attempting any heating, welding, or cutting operations.
- Never release a large amount of fuel gas, especially acetylene. Methane and propane tend to concentrate in and along low areas. Both gases can ignite at a considerable distance from the release point. Acetylene is lighter than air, but it is even more dangerous than methane. When mixed with air or oxygen, acetylene will explode at much lower concentrations than any other fuel.
- To prevent fires, maintain a neat and clean work area. Also, make sure that any metal scrap or slag is cold before disposal.

WARNING

Before welding containers, such as tanks or barrels, check to see if they have ever held any explosive, hazardous, or flammable materials. These include, but are not limited to, petroleum products, citrus products, or chemicals that decompose into toxic fumes when heated. As standard practice, always clean and then fill any tanks or barrels with water, or purge them with an appropriate purging gas to displace any oxygen.

1.1.3 Work Area Ventilation

Welders normally work within inches of their welds wearing special protective helmets. Vapors from the welds can be hazardous. The following list provides work area ventilation guidelines:

- Always follow the required confined space procedures before conducting any welding in a confined space.
- Never use oxygen for ventilation in confined spaces.

WARNING

An oxygen monitor may be required when working in a confined space.

GOING GREEN

Water Disposal

To protect the environment and save resources, make sure to properly dispose of any water used in the cutting of tanks, barrels, or various metals. If the water can be reused, save it and use it again for the next cutting.

- Always perform welding operations in a well-ventilated area. Welding operations involving zinc or cadmium materials or coatings result in toxic fumes. For long-term welding of these materials, always wear an approved, full-face SAR that uses breathing air supplied from outside of the work area. For occasional, very short-term exposure, you may use a HEPA-rated or metal-fume filter on a standard respirator.
- Make sure confined spaces are properly ventilated for welding operations.

2.0.0 WELDING EQUIPMENT SETUP

Before welding can take place, the work area must be readied, the welding equipment set up, and the metals to be welded cleaned and prepared. The following sections explain how to prepare the area and how to set up the GTAW equipment for welding aluminum plate.

2.1.0 Preparing the Welding Area

To practice welding, you will need a welding table, bench, or stand. The preferred welding surface is aluminum. Provisions must be available for mounting practice welding coupons out of position.

To set up the area for welding, follow these steps:

Step 1 Make sure that the area is properly ventilated. Use doors, windows, and fans.

Step 2 Check the area for fire hazards. Remove any flammable materials before proceeding.

Step 3 Locate the nearest fire extinguisher. Do not proceed unless the extinguisher is charged and you know how to use it.

Step 4 Set up flash shields around the welding area.

2.2.0 Selecting Aluminum Filler Metals

The aluminum filler metal typically used for the GTAW process is an uncoated solid metal filler rod. The metallurgical content of the filler metal used in GTAW is chosen for its compatibility with the base metal being welded. Filler metals are available for almost all types of aluminum.

2.2.1 Manufacturer's Classification

All filler rods made in accordance with AWS specifications must have the AWS classification number printed on the shipping container. In addition, filler rod manufacturers frequently print their own unique classification name (or number) on the shipping container. On some jobs, welders must keep proof of the filler rods they use on certain welds. One way to do this is to cut the filler rods in half and burn the rods from their cut ends. The rod ends showing the manufacturers' ID numbers can then be saved for recordkeeping.

2.2.2 Filler Rod Sizes

Aluminum filler rods come in standard lengths of 18" and 36". Common sizes (diameters) are $\frac{1}{16}$", $\frac{3}{32}$", $\frac{1}{8}$", $\frac{5}{32}$", $\frac{3}{16}$", and $\frac{1}{4}$". The rods are commonly packaged in 5-, 10-, 25-, or 50-pound packages.

2.2.3 Filler Metal Selection Considerations

Because there are so many types of filler metals available, the welder must be able to distinguish between them and select the correct filler metal for the job. There are two basic factors to consider when selecting aluminum filler metals. They are the welding procedure specification and the base metal composition.

If there is a WPS or site quality standard, the filler rod type will be specified. It will be given as an AWS classification and/or manufacturer's standard. It will sometimes include a manufacturer's name. When there is a WPS, the filler metal that is specified must be used.

CAUTION

Filler metal specifications in the WPS or site quality standards must be followed. Failure to use the specified filler metal will result in a weld quality issue.

Aluminum filler rods for welding aluminum with GTAW are usually selected for their compatibility with the chemical composition of the base metal. There are many different types of aluminum base metals and filler metals. Because of this, it is important to refer to the filler metal manufacturer's table or chart.

Hot Tip

Manufacturers' Data for Filler Metals

Most filler metal manufacturers publish paperback welding guides that include information about their filler materials. These booklets can be obtained through your distributor free or for a small charge.

Besides showing a filler metal guide, *Table 1* also includes a special notes section.

In many cases, only a few filler alloys are used to weld the majority of aluminum alloys. Filler alloys, such as 5356 and 4043, can be used for most aluminum fusion-welding applications. Always check the WPS or job standards to determine the correct filler metal for the job.

High quality aluminum welds can only be achieved if the filler metal that is used is clean and of high quality. Unclean filler metal transfers contaminants into the weld pool. Filler metal must be used as soon as possible after removal from the shipping or storage packaging. When this is not possible, the filler metal must be stored in air-tight storage containers to prevent further oxidation of the rod.

For the welding exercises in this module, use the $\frac{3}{32}$", $\frac{1}{8}$", $\frac{5}{32}$", and/or $\frac{3}{16}$" aluminum filler material classifications shown in *Table 1*, or their equivalents, to weld the aluminum plate coupons. Remove only a small number of filler rods at a time. Store the rest in the package to keep them clean. Before using a filler metal, check it for contamination, such as corrosion, dirt, oil, grease, or burned ends. Contamination will cause weld defects. Use denatured alcohol or acetone and a clean, oil-free rag to clean the filler metal. Use emery cloth or stainless steel wool to remove oxides. Snip off any burned ends after the filler metal has been cleaned. If the filler metal cannot be cleaned, do not use it. The filler metal should be cleaned with the same method that was used to clean the base metal.

Hot Tip

Filler Rod Size Selection

Use the following as a guideline for GTAW filler rod size selection: the filler rod diameter should be close to the diameter of the tungsten electrode.

Table 1 Filler Metal Guide for General Purpose Welding of Aluminum

Base Metals	319, 333 354, 355 C355	13, 43, 344 356, A356 A357, 359	214, A214 B214, F214	7039 A612, C612 D612, 7005 k	6070	6061, 6063 6101, 6151 6201, 6951	5456	5454	5154 5254 a	5086	5083	5052 5652 a	5005 5050	3004 Alc. 3004	2219	2014 2024	1100 3003 Alc. 3003	1060 EC
1060, EC	ER4145 c,i	ER4043 i,f	ER4043 e,i	ER4043 i	ER4043 i	ER4043 i	ER5356 c	ER4043 e,i	ER4043 e,i	ER5356 c	ER5356 c	ER4043 i	ER1100 c	ER4043	ER4145	ER4145	ER1100 c	ER1260 c,j
1100, 3003 Alclad 3003	ER4145 c,i	ER4043 i,f	ER4043 e,i	ER4043 i	ER4043 i	ER4043 i	ER5356 c	ER4043 e,i	ER4043 e,i	ER5356 c	ER5356 c	ER4043 e,i	ER4043 e	ER4043 e	ER4145	ER4145	ER1100 c	
2014, 2024	ER4145 g	ER4145			ER4145	ER4145									ER4145 g	ER4145 g		
2219	ER4145 g,c,i	ER4145 c,i	ER4043 i	ER4043 i	ER4043 f,i	ER4043 f,i	ER4043	ER4043 i	ER4043 i	ER4043	ER4043	ER4043 i	ER4043	ER4043	ER2319 c,f,i			
3004 Alclad 3004	ER4043 i	ER4043 i	ER5654 b	ER5356 e	ER4043 e	ER4043 b	ER5356 e	ER5654 b	ER5654 b	ER5356 e	ER5356 e	ER4043 e,i	ER4043 e	ER4043 e				
5005, 5050	ER4043 i	ER4043 i	ER5654 b	ER5356 e	ER4043 e	ER4043 b	ER5356 e	ER5654 b	ER5654 b	ER5356 e	ER5356 e	ER4043 e,i	ER4043 d,e					
5052, 5652 a	ER4043 i	ER4043 b,i	ER5654 b	ER5356 e,h	ER5356 b,c	ER5356 b,c	ER5356 b	ER5654 b	ER5654 b	ER5356 e	ER5356 e	ER5654 a,b,e						
5083		ER5356 c,e,i	ER5356 e	ER5183 e,h	ER5356 e	ER5356 e	ER5183 e	ER5356 e	ER5356 e	ER5356 e	ER5183 e							
5086		ER5356 c,e,i	ER5356 e	ER5356 e,h	ER5356 e	ER5356 e	ER5356 e	ER5356 b	ER5356 b	ER5356 e								
5154, 5254 a		ER4043 b,i	ER5654 b	ER5356 b,h	ER5356 b,c	ER5356 b,c	ER5356 b	ER5654 b	ER5654 a,b									
5454	ER4043 i	ER4043 b,i	ER5654 b	ER5356 b,h	ER5356 b,c	ER5356 b,c	ER5356 b	ER5554 c,e										
5456		ER5356 c,e,i	ER5356 e	ER5556 e,h	ER5356 e	ER5356 e	ER5556 e											
6061, 6063, 6101 6201, 6151, 6951	ER4145 c,i	ER4043 b,i	ER5356 b,c	ER5356 b,c,h,i	ER4043 b,i	ER4043 b,i												
6070	ER4145 c,i	ER4043 e,i	ER5356 c,e	ER5356 c,e,h,i	ER4043 e,i													
7039 A612, C612 D612, 7005 k	ER4043 i	ER4043 b,h,i	ER5356 b,h	ER5039 e														
214, A214 B214, F214		ER4043 b,i	ER5654 b,d															
13, 43, 344 356, A356 A357, 359	ER4145 c,i	ER4043 d,i																
319, 333 354, 355, C355	ER4145 d,c,i																	

NOTES:

1. Service conditions such as immersion in fresh or salt water, exposure to specific chemicals, or a sustained high temperature (over 150°F) may limit the choice of filler metals.
2. Recommendations in this table apply to gas shielded arc welding processes. For gas welding, only R1100, R1260, and R4043 filler metals are ordinarily used.
3. Filler metals designated with ER prefix are listed in AWS specification A5.10.
   a. Base metal alloys 5652 and 5254 are used for hydrogen peroxide service. ER5654 filler metal is used for welding both alloys for low-temperature service (150°F and below).
   b. ER5183, ER5356, ER5554, ER5556, and ER5654 may be used. In some cases they provide (1) improved color match after anodizing treatment, (2) highest weld ductility, and (3) higher weld strength. ER5554 is suitable for elevated temperature service.
   c. ER4043 may be used for some applications.
   d. Filler metal with the same analysis as the base metal is sometimes used.
   e. ER5183, ER5356, or ER5556 may be used.
   f. ER4145 may be used for some applications.
   g. ER2319 may be used for some applications.
   h. ER5039 may be used for some applications.
   i. ER4047 may be used for some applications.
   j. ER1100 may be used for some applications.
   k. This refers to 7005 extrusions only.
4. Where no filler metal is listed, the base metal combination is not recommended for welding.

402T01.EPS

2.3.0 Preparing Welding Coupons

Aluminum tends to be easily contaminated. Because of this, proper coupon preparation is very important. Aluminum coupons must be cut to the desired lengths, cleaned, and beveled to the proper angles before they are positioned for welding.

> **WARNING**
>
> Grinding wheels used on aluminum must be made for aluminum only. They should not be used on any other type of metal or on surfaces contaminated with grease or oil. Using other types of grinding wheels on aluminum may cause the grinding wheels to shatter.

2.3.1 Cutting and Cleaning Aluminum Plate Coupons

When possible, welding coupons should be cut from aluminum plate that is ⅜" thick. If this size is not readily available, aluminum plate ¼" to ¾" thick can be used. Clean oil and grease from the coupons with an approved chemical, such as acetone. After oil and grease have been removed, use a stainless steel wire brush to remove all other contaminants and oxidation. Use only light pressure.

> **CAUTION**
>
> To prevent contamination, do not use a stainless steel wire brush on anything else after it has been used on aluminum.

> **WARNING**
>
> To prevent injury when using cleaning solvents, always comply with the MSDS guidelines for the solvents being used.

2.3.2 Preparing Practice Weld Coupons

The coupons must be shaped to allow the following welds:

- *Stringer or weave beads, running beads, and overlapping beads* – The plate coupons can be any size or shape that can be easily handled.
- *Fillet welds* – Cut the metal into 4" × 5" rectangles for the base and 3" × 5" rectangles for the web (*Figure 2*).

> **GOING GREEN**
>
> **Conserve Aluminum Practice Coupons**
>
> For practice welding, aluminum is expensive and difficult to obtain. To conserve resources and reduce waste in landfills, completely use weld coupons until all surfaces have been welded upon. Cutting coupons apart and reusing the pieces conserves materials. Use material that cannot be cut into weld coupons to practice running beads.

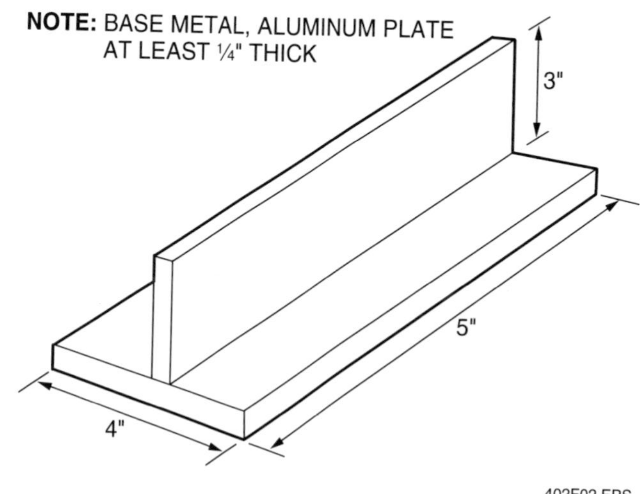

Figure 2 Fillet weld coupons.

- *V-groove welds with backing* – *Figure 3* shows a number of weld preparations and geometries. Joint G illustrates a V-groove weld with backing. As shown in *Figure 4*, cut the metal into 3" × 5" rectangles with one (or both) of the 5" lengths beveled at 30 degrees. Grind a $\frac{1}{32}$" to ⅛" root face on the bevel as directed by your instructor. Cut backing strips at least 1½" wide and 6" long from the same metal as the beveled pieces (*Figure 5*).

Follow these steps to prepare each V-groove weld coupon with backing.

> **NOTE**
>
> This module uses joint G shown in *Figure 3* for practice purposes, but the other configurations could also be used.

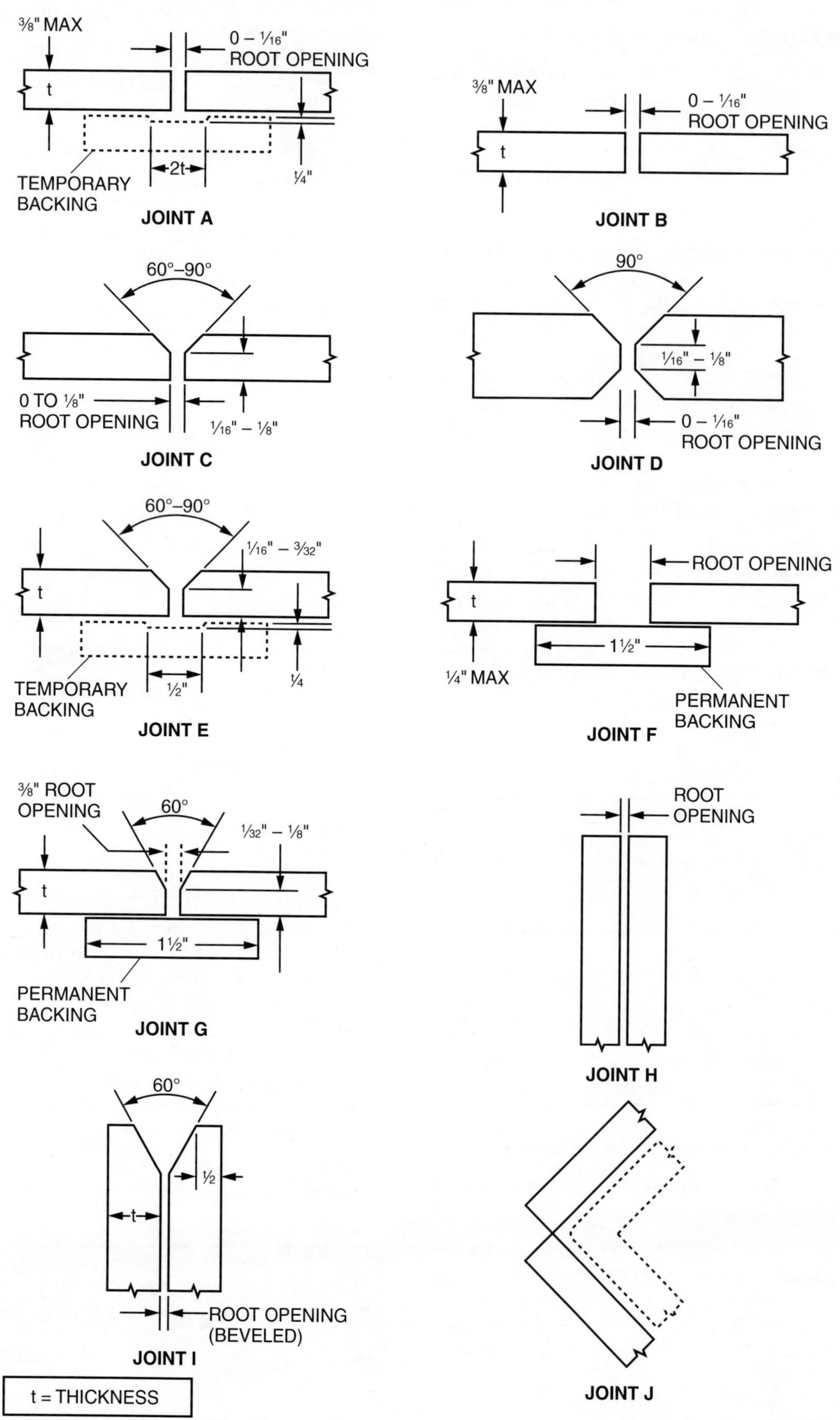

Figure 3 GTAW preparations and geometries.

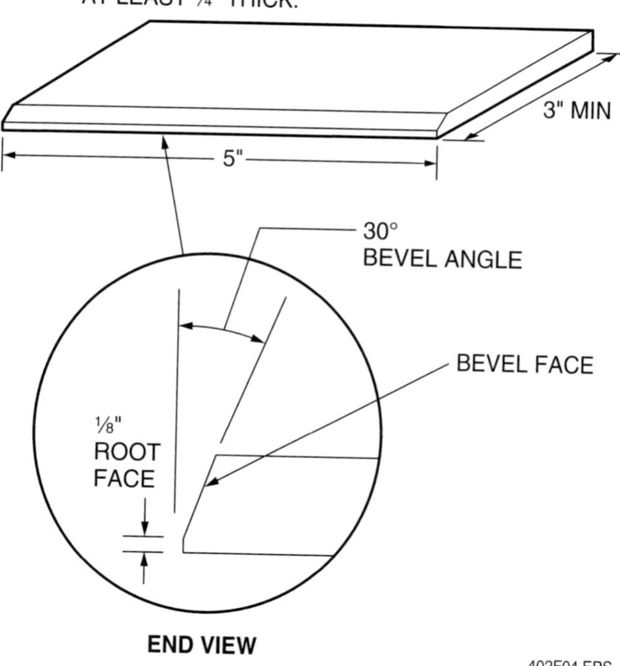

Figure 4 Preparing the weld coupon.

Step 1 Check the bevel. The bevel angle should be about 30 degrees and the root face ⅛". If the bevel has been cut thermally, machine the bevel back ⅛" before machining the root face.

Step 2 Center the beveled strips on the backing strip with a ⅜" root opening (*Figure 5*), and tack-weld them in place. If desired, use a piece of metal or rod with the required thickness as a temporary spacer. Place the tack welds on the reverse side of the joint in the lap formed by the backing strip and the beveled plate. Use three to four ½" tack welds on each beveled plate. Be certain that the backing strip is tight against the beveled plates. Make sure that the surface of the backing strip under the root opening has been cleaned of all oxide coating before tacking the strip in place.

Hot Tip

Backing Strip

When the backing strip is ½" to 1" longer than the plates at each end of the weld groove, it allows the welder to start and stop the bead outside of the weld groove.

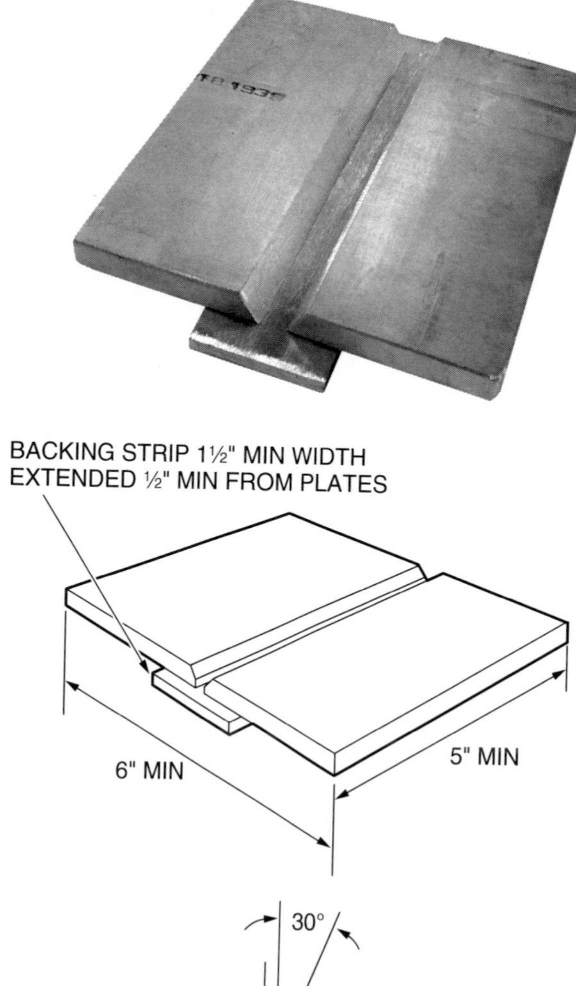

Figure 5 V-groove with metal backing weld coupon.

When welding in the horizontal position, an alternate joint preparation can be used (*Figure 6*). The alternate joint has one plate beveled at about 45 degrees and the other plate at about 15 degrees. The plates are positioned with the 45-degree beveled plate above the 15-degree plate and with a ⅜" root opening.

NOTE

Check with your instructor about whether to use the standard V-groove preparation or the alternate preparation for horizontal weld coupons.

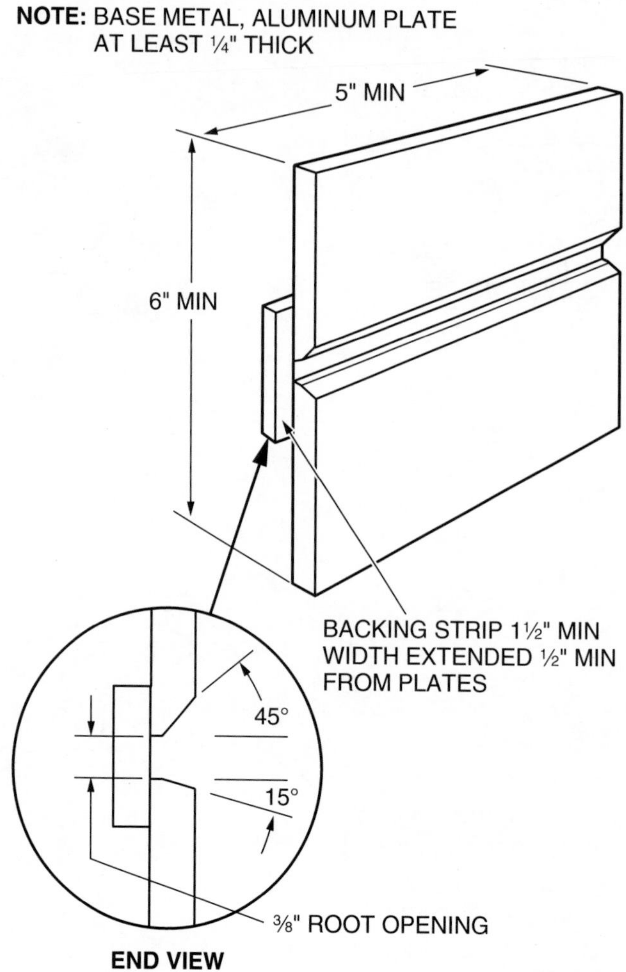

Figure 6 Alternate horizontal weld coupons.

2.3.3 Positioning Aluminum Plate Coupons for Welding

Different welding techniques are used in different welding situations. This section explains the various positions in which the aluminum plate coupons can be placed. It also covers the requirements for each respective position. GTAW welding of aluminum plate requires specific preparation and shapes. Refer to *Figure 3* for different joint preparations and geometries.

The specifications for each weld are determined by a number of factors. *Tables 2, 3,* and *4* show the specifications for groove welds, fillet and lap welds, and edge and corner welds, respectively.

2.4.0 Selecting Shielding Gas

Shielding gas is selected according to the material being welded. Argon is the most commonly used shielding gas. Argon can be used for welding a wide range of materials, including aluminum, titanium, steel, and stainless steel.

Helium and helium/argon mixtures raise the temperature of the arc. This rise in temperature promotes higher welding speeds and deeper weld penetration. The disadvantages of using helium or a **helium/argon mixture** include difficulty in starting the arc and the high cost of the gas.

Table 2 Specifications for Making GTAW Groove Welds on Aluminum Plate without Backing

Aluminum Thickness (inch)	Weld Position[1]	Edge Prep[2] (joint)	Root Opening[3] (inch)	Preheat[4] (°F)	Weld Passes	Filler Diameter (inch)	Tungsten Electrode Diameter (inch)	Gas Cup Inside Diameter (inch)	Argon (cfh)[5]	AC (amps)[5]	Arc Travel Speed (ipm)[5]	Approx. Filler Rod Consumption (lb/100 ft)[5]
1/16	F, V, H	A or B	1/16	None	1	3/32	1/16–3/32	3/8	20	70–100	8–10	0.5
	O	A or B	1/16	None	1	3/32	1/16	3/8	25	60–75	8–10	0.5
3/32	F	A or B	3/32	None	1	1/8	3/32–1/8	3/8	20	95–115	8–10	1
	V, H	A or B	3/32	None	1	3/32–1/8	3/32	3/8	20	85–110	8–10	1
	O	A or B	3/32	None	1	3/32–1/8	3/32–1/8	3/8	25	90–110	8–10	1
1/8	F	A or B	1/8	None	1–2	1/8–5/32	1/8	7/16	20	125–150	10–12	2
	V, H	A or B	3/32	None	1–2	1/8	1/8	7/16	20	110–140	10	2
	O	A or B	3/32	None	1–2	1/8–5/32	1/8	7/16	25	115–140	10–12	2
3/16	F	D-60°	1/8	None	2	5/32–3/16	5/32–3/16	7/16–1/2	25	170–190	10–12	4.5
	V	D-60°	3/32	None	2	5/32	5/32	7/16	25	160–175	10–12	4.5
	H	D-90°	3/32	None	2	5/32	5/32	7/16	25	155–170	10–12	5
	O	D-110°	3/32	None	2	5/32	5/32	7/16	30	165–180	10–12	6
1/4	F	D-60°	1/8	None	2	3/16	3/16–1/4	1/2	30	220–275	8–10	8
	V	D-60°	3/32	None	2	3/16	3/16	1/2	30	200–240	8–10	8
	H	D-90°	3/32	None	2–3	5/32–3/16	5/32–3/16	1/2	30	190–225	8–10	9
	O	D-110°	3/32	None	2	3/16	3/16	1/2	35	210–250	8–10	10
3/8	F	D-60°	1/8	Optional up to 250°F maximum	2	3/16–1/4	1/4	5/8	35	315–375	8–10	15.5
	F	E	3/32		2	3/16–1/4	1/4	5/8	35	340–380	8–10	14
	V	D-60°	3/32		3	3/16	3/16–1/4	5/8	35	260–300	8–10	19
	V, H, O	E	3/32		2	3/16	3/16–1/4	5/8	35	240–300	8–10	17
	H	D-90°	3/32		3	3/16	3/16–1/4	5/8	35	240–300	8–10	22
	O	D-110°	3/32		3	3/16	3/16–1/4	5/8	40	260–300	8–10	32

1. F = flat; V = vertical; H = horizontal; O = overhead
2. Refer to Figure 3 for edge preparation and joint specifications.
3. Always check the WPS or site-specific specifications.
4. Preheating at excessive temperatures or for extended periods of time will reduce weld strength. This is particularly true for alloys in heat-treated tempers.
5. cfh = cubic feet per hour; amps = amperes; ipm = inches per minutes; lb/100 ft = pounds per hundred feet

402T02.EPS

Impurities in the shielding gas or mixture with other gases, such as carbon dioxide, will contaminate the weld and change the color of the aluminum, usually to black.

CAUTION

Always use the correct gas when performing GTAW. Accidental use of a GMAW gas mix can cause rapid destruction of the tungsten and contamination of the weld.

2.5.0 Welding Equipment

Identify a proper welding machine for GTAW. Follow these steps to set the machine up for use:

Step 1 Verify that the welding machine can be used for GTAW with or without internal gas shielding control. If desired, identify an optional cooling unit.

Step 2 Identify an air-cooled or water-cooled GTAW torch. Make sure that it is compatible with the welding machine and any optional cooling unit.

Table 3 Specifications for Making GTAW Fillet and Lap Welds on Aluminum Plate

Aluminum Thickness (inch)	Weld Position[1]	Preheat[2] (°F)	Weld Passes	Filler Diameter (inch)	Pure Tungsten Electrode Diameter (inch)	Gas Cup Inside Diameter (inch)	Argon Flow (cfh)	AC (amps)	Arc Travel Speed (ipm)	Approximate Filler Rod Consumption (lb/100 ft)
1/16	F, V, H	None	1	3/32	1/16–3/32	3/8	16	70–100	8–10	0.5
	O	None	1	3/32	1/16–3/32	3/8	20	65–90	8–10	0.5
3/32	F	None	1	3/32–1/8	1/8–5/32	3/8	18	110–145	8–10	1
	V, H	None	1	3/32	3/32–1/8	3/8	18	90–125	8–10	1
	O	None	1	3/32	3/32–1/8	3/8	20	110–135	8–10	1
1/8	F	None	1	1/8	1/8–5/32	7/16	20	135–175	10–12	2
	V, H	None	1	1/8	3/32–1/8	3/8	20	115–145	8–10	2.5
	O	None	1	1/8	3/32–1/8	7/16	25	125–155	8–10	2
3/16	F	None	1	5/32	5/32–3/16	1/2	25	190–245	8–10	4.5
	V, H	None	1	5/32	5/32–3/16	1/2	25	175–210	8–10	5.5
	O	None	1	5/32	5/32–3/16	1/2	30	185–225	8–10	4.5
1/4	F	None	1	3/16	3/16–1/4	1/2	30	240–295	8–10	7
	V, H	None	1	3/16	3/16	1/2	30	220–265	8–10	9
	O	None	1	3/16	3/16	1/2	35	230–275	8–10	7
3/8	F	Optional	2	3/16	1/4	5/8	35	325–375	8–10	17
	V	up to	2	3/16	3/16–1/4	5/8	35	280–315	8–10	20
	H	250°F	3	3/16	3/16–1/4	5/8	35	270–300	8–10	20
	O	maximum	3	3/16	3/16–1/4	5/8	40	290–335	8–10	17

1. F = flat; V = vertical; H = horizontal; O = overhead
2. Preheating at excessive temperatures or for extended periods of time will reduce weld strength. This is particularly true for alloys in heat-treated tempers.

402T03.EPS

Table 4 Specifications for Making GTAW Edge and Corner Welds on Aluminum Plate

Aluminum Thickness (inch)	Edge Prep	Weld Passes	Filler Rod Diameter (inch)	Tungsten Electrode Diameter (inch)	Gas Cup Inside Diameter (inch)	Argon Flow (cfh)	AC[1,2] (amps)	Arc Travel Speed (ipm)	Approximate Filler Rod Consumption (lb/100 ft)
1/16	I,K	1	3/32	1/16	3/8	20	60-85	10-16	0.5
3/32	I,K	1	1/8	3/32	3/8	20	90-120	10-16	1
1/8	I,K	1	1/8-5/32	1/8	3/8	20	115-150	10-16	2
3/16	J,K	1	5/32	5/32	7/16	25	160-220	10-16	4.5
1/4	J,K	2	3/16	3/16	1/2	30	200-250	8-12	7

1. Higher currents and welding speeds can be employed if a temporary backing is used for corner joints.
2. Use the low side of the current range for horizontal and vertical welds.

402T04.EPS

Step 3 Know the location of the primary disconnect.

Step 4 Configure the welding machine for aluminum GTAW welding (*Figure 7*) as directed by your instructor. Configure the torch polarity to AC. Equip the torch with a properly prepared 3/32" or 1/8" EWP or EWZr tungsten electrode.

NOTE

Pure tungsten (EWP) electrodes are used for welding aluminum and magnesium with AC and high frequency. Zirconiated tungsten (EWZr) electrodes are used for welding aluminum and magnesium with AC and high frequency for welds in which tungsten inclusions are not tolerated and higher current capacity is desired.

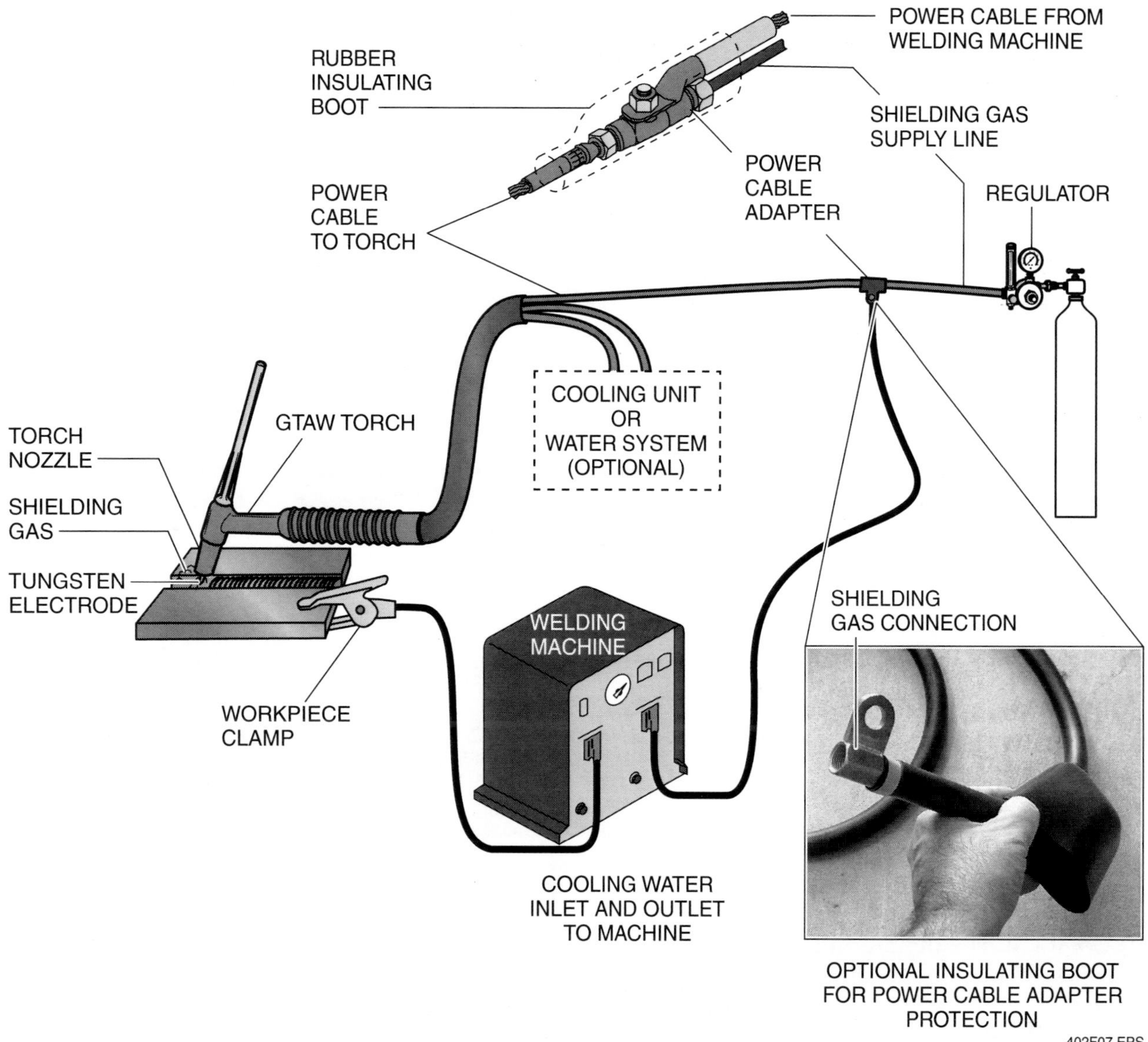

Figure 7 Configuration diagram of typical GTAW welding equipment.

Step 5 Connect the proper argon or argon/helium shielding gas for the application as directed by your instructor, filler metal manufacturer, WPS, or site quality standards.

Step 6 Connect the clamp of the workpiece lead to the workpiece.

Step 7 Turn on the welding machine. Adjust the flowmeter and purge the torch according to the manufacturer's instructions.

Step 8 Set the initial welding current for the GTAW application.

3.0.0 GTAW Techniques

GTAW weld bead characteristics and quality are affected by several factors that are influenced by the way the welder handles the torch. These factors include the following:

- Torch travel speed and arc length
- Torch angles
- Torch and filler-metal handling techniques

3.1.0 Torch Travel Speed and Arc Length

Torch travel speed and arc length affect the GTAW weld puddle and the penetration of the weld. Slower travel speeds allow more heat to

Going Green: Advanced Inverter Power Sources

An advanced inverter power source can make welding aluminum much easier, while saving electrical energy. With features such as HF start only, current pulsing, AC frequency adjustment, and weld current sequencing (start/stop ramping), these machines provide additional control over some of the variables that affect the quality of a weld. These machines also allow the use of tapered and blunted 2 percent thoriated or ceriated tungsten electrodes for aluminum welding in an AC mode as well as a DC mode. With a tapered electrode, an operator can place a ⅛" bead on a fillet weld made from ⅛" plate. When a normal balled electrode is used, it makes a much larger bead that has to be ground down to final size.

402SA01.EPS

concentrate and form larger, more deeply penetrating puddles. Faster travel speeds prevent heat buildup and form smaller, shallower puddles. Arc length is the major control for bead width. As the torch is raised, arc length increases, voltage increases, and the bead width increases.

3.2.0 Torch Angles

The two basic torch angles that must be controlled when performing GTAW are the work angle and the travel angle. The definition of these angles is the same as for all other methods of welding.

3.2.1 Work Angle

The torch work angle (*Figure 8*) is an angle less than 90 degrees between a line perpendicular to the major workpiece surface at the point of electrode contact and a plane determined by the electrode axis and the weld axis. For a T-joint or corner joint, the line is perpendicular to the nonbutting member. For pipe, the plane is determined by the electrode axis and a line tangent to the pipe surface at the same point.

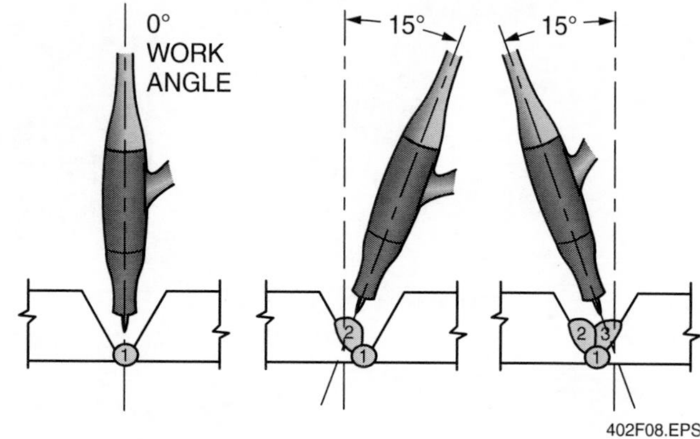

Figure 8 Typical torch work angles.

3.2.2 Travel Angle

The torch travel angle (*Figure 9*) is an angle less than 90 degrees between the electrode axis and a line perpendicular to the weld axis at the point of electrode contact in a plane determined by the electrode axis and the weld axis. For pipe, the plane is determined by the electrode axis and a line tangent to the pipe's surface at the same point.

A push angle is used for GTAW. A push angle is created when the torch is tilted back so that the electrode tip precedes the torch in the direction of the weld. In this position, the electrode tip and shielding gas are being directed ahead of the weld bead. This provides a cleaning action of the positive portion of the AC voltage cycle. Push angles of 10 to 15 degrees are normally used for GTAW of aluminum plate.

Hot Tip

Excessive Push Angle

Excessive push angles tend to draw air from under the back edge of the torch nozzle where the air mixes with the shielding gas stream and contaminates the weld.

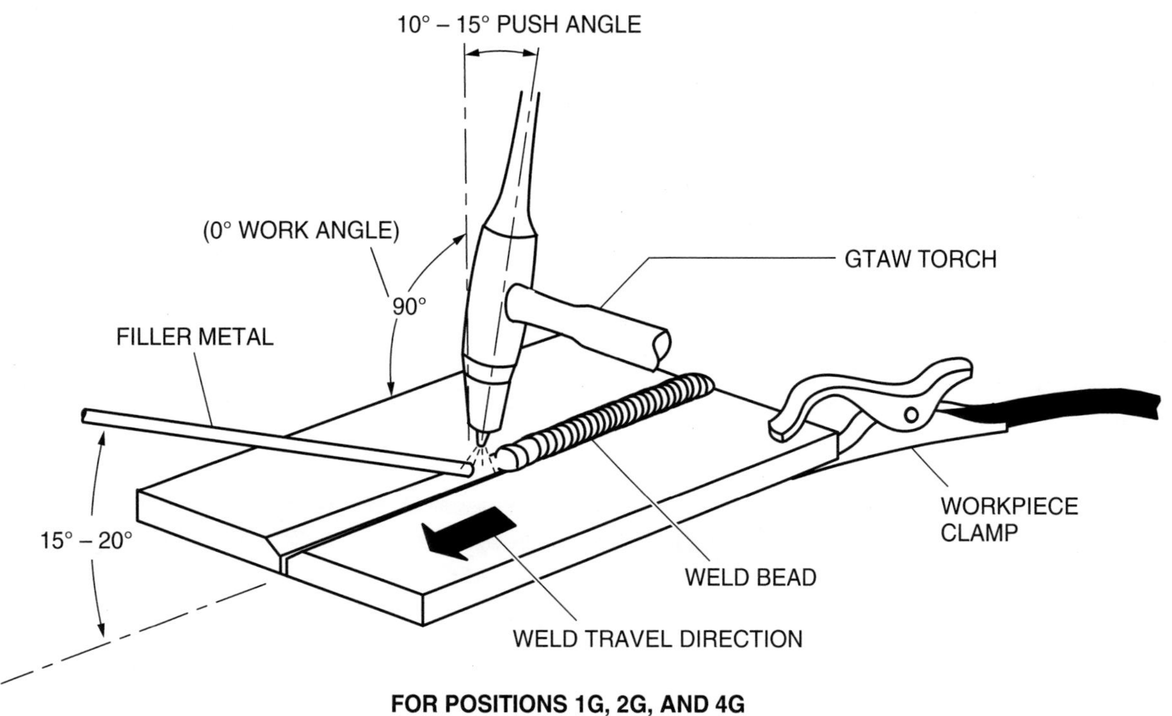

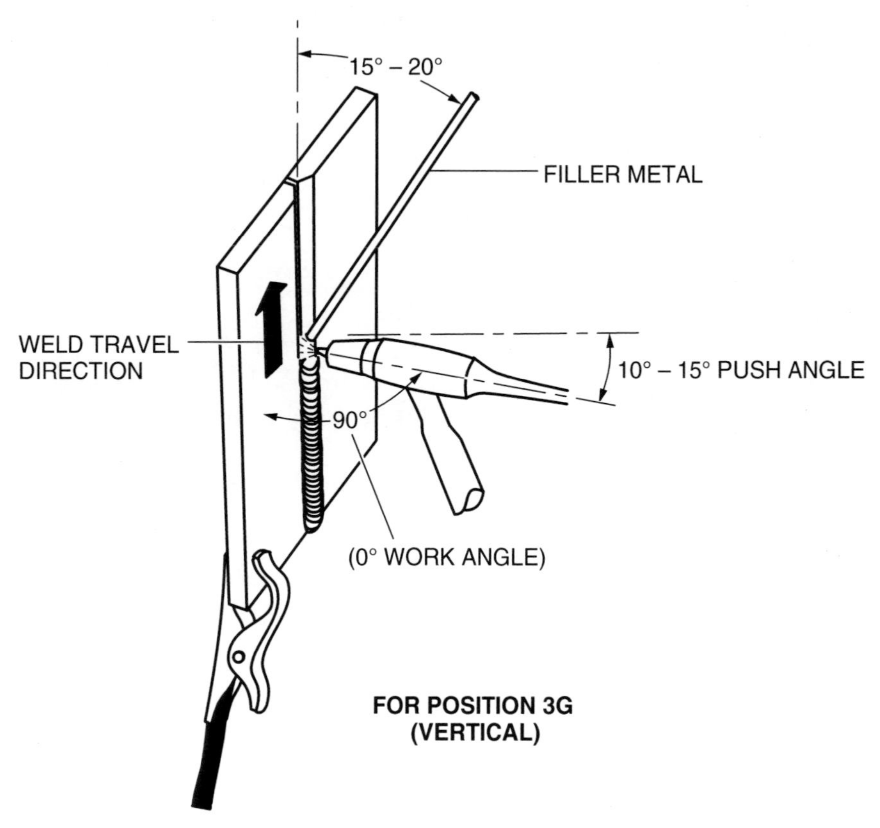

Figure 9 Torch travel angles.

3.3.0 Torch and Filler Metal Handling Techniques

The two basic torch and filler metal handling methods used to perform GTAW are the freehand and the walking-the-cup techniques. Try both, and use the one that gives you the best results.

3.3.1 Freehand Technique

The freehand technique is the preferred method for aluminum. When this method is used, the torch electrode tip is held just above the weld puddle or base metal (*Figure 10*). The torch is supported by the welder's hand like a pencil. The hand is usually steadied by resting some part of it on or against the base metal. This helps to keep the proper arc length (distance from the electrode tip to the weld pool). If required, the welder can move the torch tip in a small circular motion within the molten puddle. This maintains the puddle size and advances the puddle. Filler metal is added as needed.

The GTAW filler metal is held in the hand that is not supporting the torch. Hold the filler rod at an angle of about 20 degrees above the base metal surface and in line with the weld. Always keep the tip of the filler rod within the shielding gas envelope to protect it from atmospheric contamination and to keep it preheated.

Dab the filler metal into the leading edge of the weld puddle, using extreme care not to touch the tungsten electrode with the end of the filler metal. If the tungsten electrode touches the filler metal or weld puddle, it will become contaminated with filler metal. The electrode must then be removed, the contaminated portion broken off, and the end reformed before welding can continue. To prevent contamination of the electrode by the filler metal, move the electrode to the back edge of the weld puddle as the filler metal is dabbed into the weld puddle's leading edge.

3.3.2 Walking-the-Cup Technique

In the walking-the-cup method, the welder rests the edge of the torch nozzle (cup) against the base metal or groove edges to steady the torch and maintain a constant arc length. The torch is rocked from side to side on the edge of the cup as it is advanced. This action maintains the puddle size and heats both sides of the groove. The filler metal is added in the same manner used in the freehand technique, taking care not to contaminate the electrode.

Hot Tip

Walking-the-Cup Technique on Aluminum

The walking-the-cup technique is very difficult to use when welding aluminum. The cup tends to stick to the freshly welded metal.

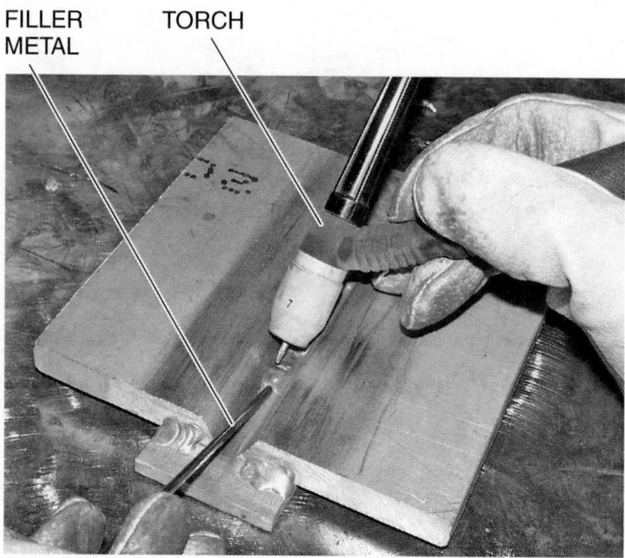

Figure 10 Freehand technique.

4.0.0 BEAD TYPES

The two basic bead types used in GTAW are weave beads and stringer beads. Stringer and weave beads are used to make welds, while overlapping stringer and weave beads are used to build pads.

4.1.0 Practicing Weave Beads

Weave beads (*Figure 11*) are made by working the weld puddle back and forth across the axis of the weld. This produces a weld bead much wider than the puddle width. The width of a weave bead is determined by the amount of oscillation or cross motion. Weave beads put more heat into the base metal.

NOTE

Always check the WPS or site quality standards to determine whether stringer or weave beads should be used.

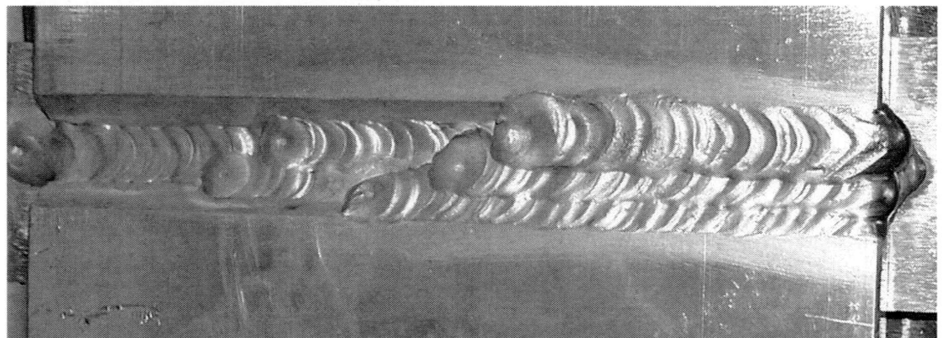

Figure 11 Weave bead.

When making a weave bead, take special care at the toes to ensure proper fusion to the base metal. Do this by slowing down or pausing slightly at the edges. Pausing at the edges will also flatten out the weld and give it the proper profile.

Practice running weave beads in the flat position. Experiment with different weave motions and push angles.

> **CAUTION**
>
> Do not exceed the recommended weave bead width. The width of weave beads is often specified in the welding code or WPS used at the site.

Follow these steps to run weave beads:

Step 1 Hold the torch at the desired angle with the electrode tip directly over the point where the weld will begin.

Step 2 Lower the torch and filler rod to the weld.

Step 3 Hold the arc in place until the weld puddle begins to form. Add filler metal.

Step 4 Slowly advance the arc in a weaving motion while maintaining the torch angle and adding filler metal (*Figure 12*).

Step 5 Continue to weld until a bead about 2" to 3" long is formed.

Step 6 Stop, and then fill the crater while slowly reducing amperage with a current control to allow the weld puddle to solidify. Turn off the current. Hold the filler rod and torch in place until the tungsten electrode and weld cool.

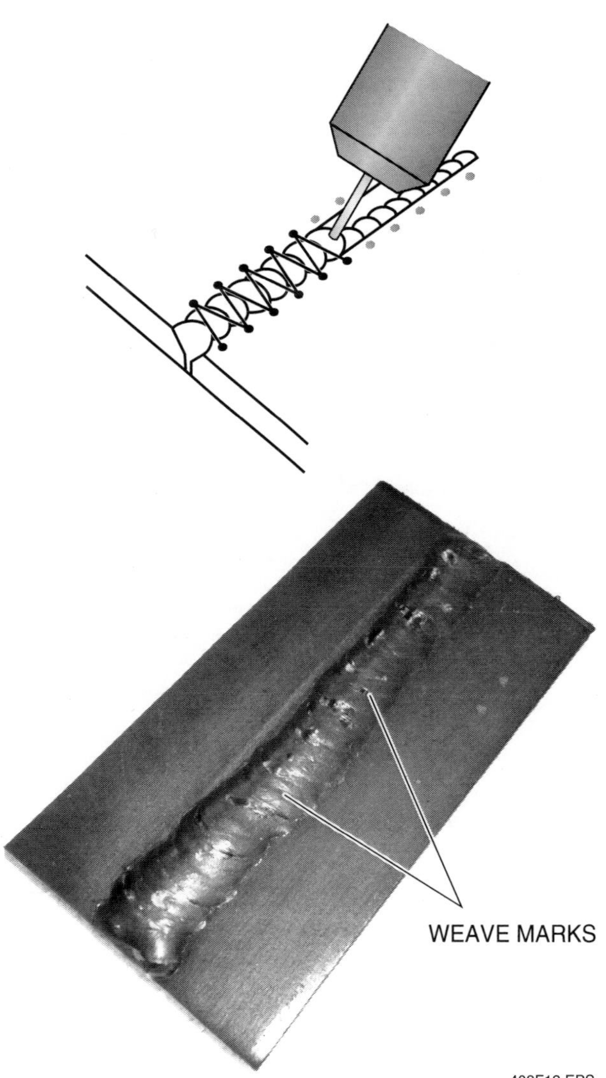

Figure 12 Weave motion.

Step 7 Inspect the bead for the following:

- Straightness of the bead
- Uniform appearance of the bead face
- Smooth, flat transition with complete fusion at the toes of the weld
- No porosity
- No excessive undercut
- Crater filled
- No cracks

Step 8 Continue practicing weave beads until you can make acceptable welds every time.

WARNING

Use pliers to handle the hot practice coupons. Wear gloves when placing the practice coupons in water. Steam will rise off the coupons and can burn or scald unprotected hands.

4.2.0 Weld Restarts

A restart is the junction where a new weld connects to the bead of a previous weld to continue the weld. A restart must be made so that it blends smoothly with the previous weld and does not stand out. An improperly made restart will create a weld defect. Whenever possible, avoid restarts by running a bead the full length of the weld joint.

Follow these steps to make a restart:

Step 1 Clean the area of the restart using an approved cleaning method.

Step 2 Hold the torch at the proper angle and arc distance. Restart the arc directly over the center of the crater.

NOTE

Welding codes do not allow arc strikes outside of the area that is to be welded.

Step 3 Move the electrode tip in a small circular motion over the crater, and add filler metal when the molten puddle is the same size as the crater.

Step 4 As soon as the puddle fills the crater, advance the puddle slightly. Continue to add filler metal as needed.

Step 5 Inspect the restart.

NOTE

A properly made restart blends into the bead and is hard to detect. If the restart has undercut, not enough time was spent in the crater to fill it. If the undercut is on one side or the other, use more of a side-to-side motion as you move back into the crater. If the restart has a lump, it was overfilled. Too much time was spent in the crater before the forward motion was resumed.

Step 6 Continue to practice restarts until they are correct.

NOTE

Use the same technique for making restarts whenever performing GTAW.

4.3.0 Weld Terminations

A weld termination is made at the end of a weld, and it normally leaves a crater. Most welding codes require the crater to be filled to the full cross section of the weld to prevent crater cracking. Filling to the full cross section of the weld can be difficult. Most terminations are at the edge of a plate where welding heat tends to build up. This makes it harder to fill the crater. Filling the crater is easier when remote control equipment with a potentiometer is used. The potentiometer allows the welding current to be reduced as the end of the weld is approached.

Follow these steps to make a termination using remote control equipment:

Step 1 As you approach the end of the weld, slowly reduce the welding current. The amount of reduction in welding current is determined by the width of the weld puddle. If the puddle begins to widen, there is too much heat. If the puddle starts to become narrower, there is not enough heat.

Hot Tip

Cooling Practice Coupons

If the practice coupon gets too hot between passes, cool it in water. Cooling with water is only done with practice coupons. Never use water to cool test coupons or any other weld. Cooling with water can cause weld cracks and affect the mechanical properties of the base metal.

Step 2 Continue to reduce the welding current while adding filler metal until the crater has been filled.

Step 3 Stop the arc and hold the torch in place with the gas postflow continuing to protect the weld metal and tungsten electrode.

Follow these steps to make a weld termination without remote control equipment:

Step 1 As you approach the end of the weld, add filler metal at a faster rate. The filler metal will absorb excess heat.

Step 2 Continue to add filler metal until the crater has been filled, and then stop the arc.

Step 3 Hold the torch in place with the gas postflow continuing to protect the weld metal, filler metal, and tungsten electrode.

CAUTION

Do not remove the torch or filler metal from the torch shielding gas flow until the puddle has solidified and cooled enough to be unaffected by the atmosphere. The gas postflow that continues after the welding has stopped protects the molten metal and the tungsten electrode. If the torch and shielding gas are removed before the weld has solidified, crater porosity or cracks can occur. Also, the filler metal will be contaminated, and the contaminated end will have to be discarded. Finally, the tungsten electrode can oxidize (turn a bluish-black color), which requires reballing or repointing the electrode.

Hot Tip

Advanced GTAW Machines with Current Ramping

Advanced GTAW machines can be programmed so that their current is automatically ramped up for the start of a weld and ramped down at the end of the weld. These advanced machines simplify weld starts and terminations.

Step 4 Inspect the weld termination for the following:

- Crater filled to the full cross section of the weld
- No crater cracks
- No crater porosity
- No excessive undercut

4.4.0 Overlapping Beads

Overlapping beads are made by depositing connective weld beads parallel to one another. Each successive bead overlaps the toe of the previous bead to form a flat surface. This is also called padding. Overlapping beads are used to build up surfaces (padding) and to make multiple-pass welds. Both stringer and weave beads can be overlapped. Properly overlapped beads, when viewed from their ends, form a nearly flat surface (*Figure 13*).

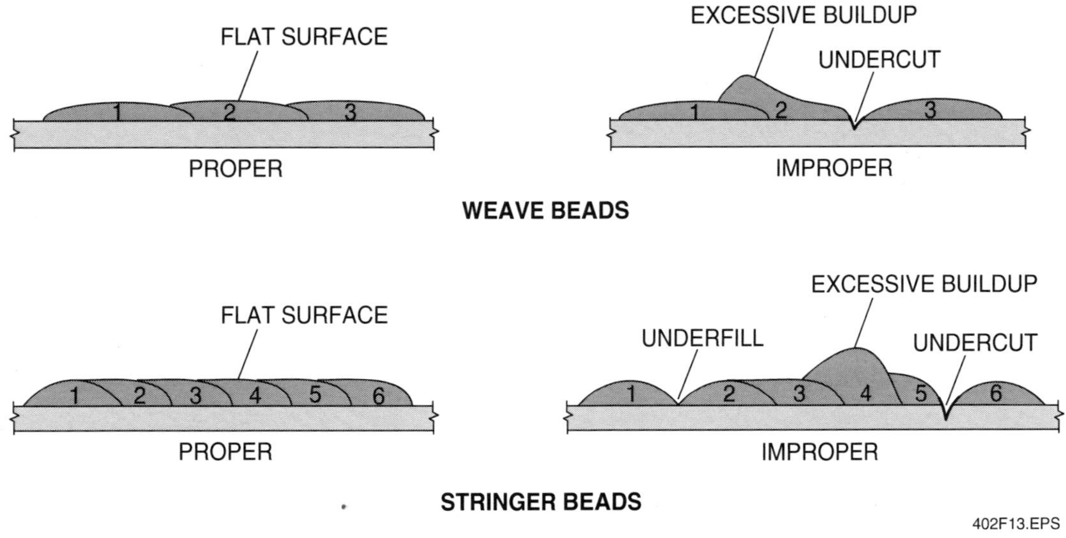

Figure 13 Proper and improper overlapping beads.

> **Hot Tip**
>
> **Thin Materials**
>
> On thin materials, stop about 1" or 2" from the edge without terminating the weld. Then backweld from the edge to the crater, and fill the crater.

Follow these steps to make GTAW overlapping stringer beads using ⅛" aluminum filler metal:

Step 1 Mark out a square on a piece of aluminum.

Step 2 Deoxidize the base metal.

Step 3 Weld a stringer bead along one edge.

Step 4 Clean the bead with a stainless steel brush.

> **CAUTION**
>
> Between weld beads, always clean the weld with a stainless steel wire brush.

Step 5 Run a second bead along the previous bead. Be sure to overlap the previous bead to obtain a good fusion and produce a flat profile.

Step 6 Continue running overlapping stringer beads until the face of the square piece of metal has been covered.

Step 7 Continue building layers of stringer beads (pads), one on top of the other, until your technique has been perfected.

4.5.0 Building a Pad with Stringer and Weave Beads

Continue to weld overlapping beads using ⅛" filler rods. Build a pad using stringer and weave beads (*Figure 14*).

402F14.EPS

Figure 14 Pad built with stringer and weave beads.

5.0.0 Fillet Welds

Fillet welds require little base metal preparation, except for the cleaning of the weld area and the removal of any excess material from cut surfaces. Any dross from cutting will cause porosity in the weld. For this reason, the codes require that this material is entirely removed prior to welding.

The most common fillet welds are made in lap and T-joints. The weld position for plate is determined by the axis of the weld and the orientation of the workpiece. The positions for fillet welding (*Figure 15*) are flat (1F), horizontal (2F), vertical (3F), and overhead (4F). In the 1F and 2F positions, the weld axis can be inclined up to 15 degrees. Any weld axis inclination for the other positions varies with the rotational position of the weld face as specified in the AWS Standards.

Fillet welds can be concave or convex, depending on the WPS or site quality standards. The welding codes require a fillet weld to have a uniform concave or convex face. However, a slightly nonuniform face is acceptable. The convexity of a fillet weld or individual surface bead must not exceed that permitted by the applicable code or standard. A fillet weld with profile defects is unacceptable and must be repaired (*Figure 16*).

> **Hot Tip**
>
> **Tacking and Aligning Workpieces**
>
> When tacking workpieces together, position them to minimize distortion when the final welds are made. To achieve this, both sides of the workpieces are usually tacked with welds that are about ½" long. After the first tack weld, use a hammer or other tool to align the workpieces side to side and end to end. Then tack the opposite side. Tack the far ends of the workpieces in the same way. Intermediate tack welds can be made every 5" to 6" as needed to minimize lengthwise distortion.

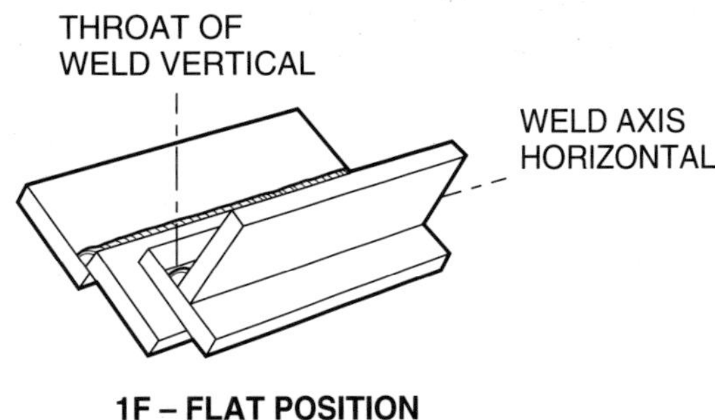

1F – FLAT POSITION

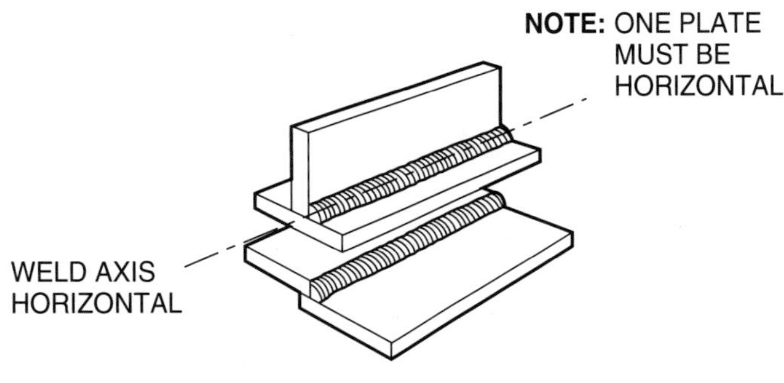

2F – HORIZONTAL POSITION

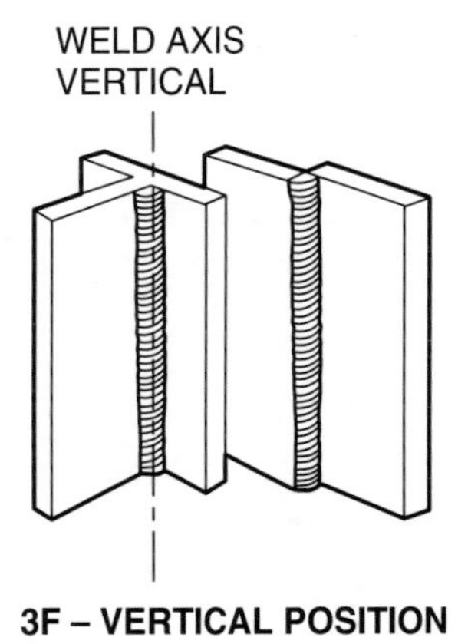

3F – VERTICAL POSITION

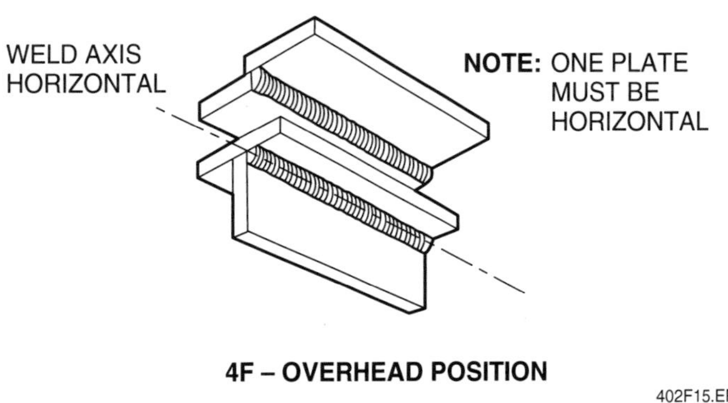

4F – OVERHEAD POSITION

402F15.EPS

Figure 15 Fillet welding positions.

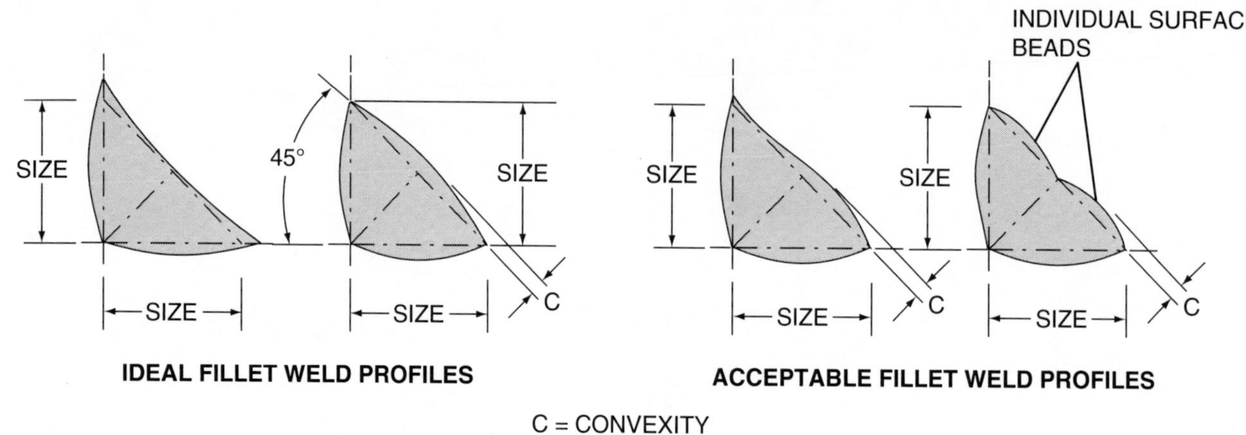

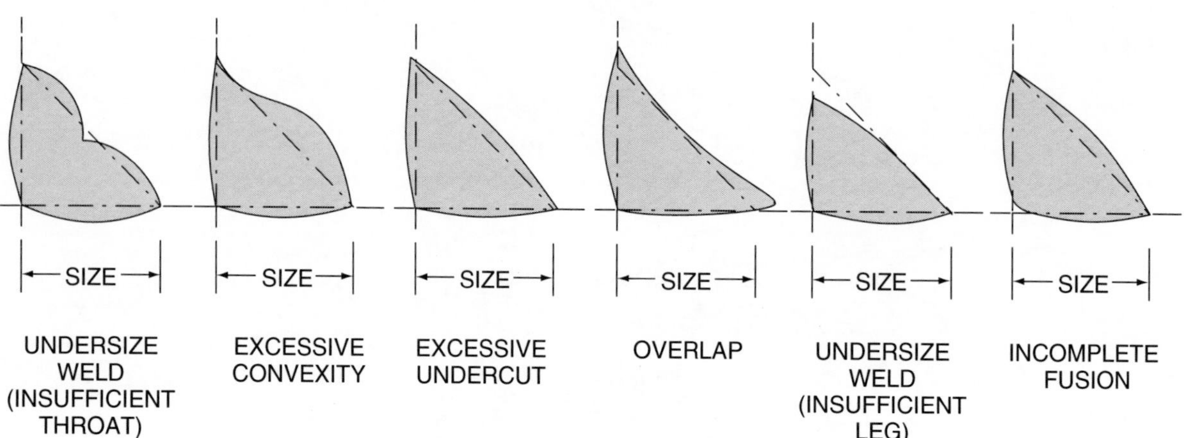

Figure 16 Ideal, acceptable, and unacceptable fillet weld profiles.

Hot Tip

T-Joint Heat Dissipation

In T-joints, the welding heat dissipates more rapidly in the thicker or nonbutting member. On various bead passes, the arc may have to be slightly more concentrated on the thicker or the nonbutting member to compensate for heat loss.

5.1.0 Practicing Flat (1F) Position Fillet Welds

Practice flat (1F) position fillet welds. Make multiple-pass (six-pass) convex fillet welds in a T-joint using an appropriate aluminum filler metal. When making flat fillet welds, pay close attention to the torch angle and travel speed. For the first bead, the torch angle is vertical (45 degrees to both plate surfaces). The angle is adjusted for all subsequent beads. Clean each completed bead with a stainless steel wire brush before starting the next bead.

Figure 17 shows the torch angles for fillet welds in the 1F position.

Follow these steps to make a flat fillet weld:

NOTE

When performing the following steps, always clean all weld beads before beginning the next bead.

Step 1 Tack two plates together to form a T-joint for the fillet weld coupon (*Figure 18*). Clean the tack welds.

Step 2 Clamp or tack-weld the coupon in the 1F position (flat).

Step 3 Run the first bead along the root of the joint. Use a work angle of 45 degrees and a 10- to 15-degree push angle.

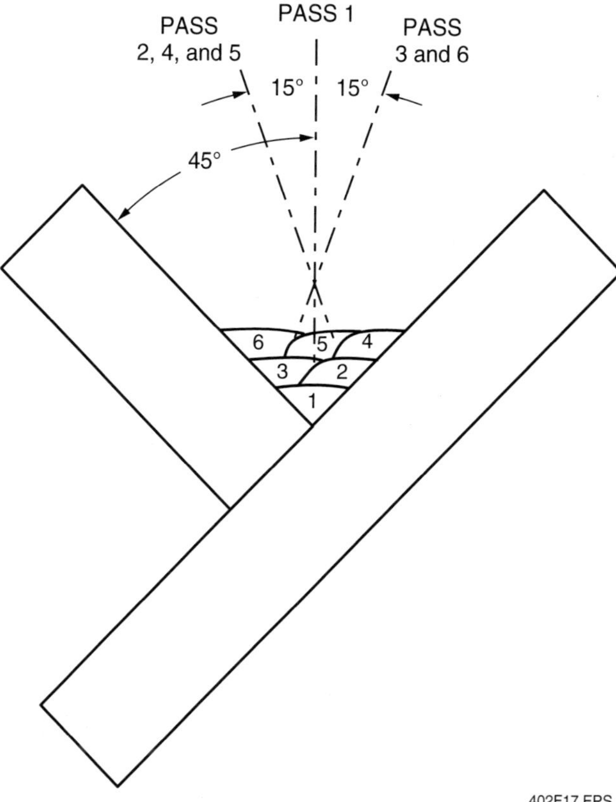

Figure 17 Multiple-pass 1F bead sequence and work angles.

Figure 18 Fillet weld coupon.

Step 4 Run the second bead along a toe of the first weld, overlapping about 75 percent of the first bead. Alter the work angle as shown in *Figure 17*. Use a 10- to 15-degree push angle with a slight oscillation.

Step 5 Run the third bead along the other toe of the first weld. Be sure to fill the groove created when the second bead was run. Use the work angle shown in *Figure 17* and a 10- to 15-degree push angle with a slight oscillation.

Step 6 Run the fourth bead along the outside toe of the second weld, overlapping about half of the second bead. Use the work angle shown in *Figure 17* and a 10- to 15-degree push angle with a slight oscillation.

Step 7 Run the fifth bead along the inside toe of the fourth weld, overlapping about half of the fourth bead. Use the work angle shown in *Figure 17* and a 10- to 15-degree push angle with a slight oscillation.

Step 8 Run the sixth bead along the toe of the fifth weld. Be sure to fill the groove created when the fifth bead was run. Use the work angle shown in *Figure 17* and a 10- to 15-degree push angle with a slight oscillation.

Step 9 Have your instructor inspect the weld. The weld is acceptable if it has the following features:

- Uniform appearance on the bead face
- Craters and restarts filled to the full cross section of the weld
- Uniform weld size of ±1⁄16"
- Acceptable weld profile in accordance with the applicable code or standard
- Smooth transition with complete fusion at the toes of the weld
- No porosity
- No excessive undercut
- No overlap
- No inclusions
- No cracks

5.2.0 Practicing Horizontal (2F) Position Fillet Welds

Practice horizontal (2F) fillet welding by placing multiple-pass fillet welds in a T-joint. Use an appropriate filler metal as directed by your instructor. When making horizontal fillet welds, pay close attention to the torch angles. For the first bead, the electrode work angle is 45 degrees. The work angle is adjusted for all other welds.

Follow these steps to make a horizontal fillet weld:

Step 1 Tack two plates together to form a T-joint for the fillet weld coupon. Clean the tack welds.

Step 2 Clamp or tack-weld the coupon in the horizontal position.

> **Hot Tip**
>
> **Raising the Working Height of a GTAW Torch**
>
> In some cases, raising the torch height (within limits) can be used to widen the second and subsequent weld beads without oscillating the torch.

Step 3 Run the first bead along the root of the joint. Use a work angle of about 45 degrees and a 10- to 15-degree push angle (*Figure 19*).

Step 4 Clean the weld.

Step 5 Run the remaining passes at the proper work angles using a 10- to 15-degree push angle and a slight oscillation (*Figure 19*). Overlap each previous pass. Clean the weld after each pass.

Step 6 Have your instructor inspect the weld. The weld is acceptable if it has the following features:

- Uniform appearance on the bead face
- Craters and restarts filled to the full cross section of the weld
- Uniform weld size of ±1⁄16"
- Acceptable weld profile in accordance with the applicable code or standard
- Smooth transition with complete fusion at the toes of the weld
- No porosity
- No excessive undercut
- No overlap
- No inclusions
- No cracks

5.3.0 Practicing Vertical (3F) Position Fillet Welds

Practice vertical (3F) fillet welding by placing multiple-pass fillet welds in a T-joint. Use an appropriate filler metal as directed by your instructor. Vertical welds are usually made by welding uphill from the bottom to the top using a torch push angle (up-angle). Because of the uphill welding and push angle, this type of weld is sometimes called vertical-up fillet welding. Either stringer or weave beads can be used for vertical welding. On the job, the site WPS or site quality standard will specify which technique to use.

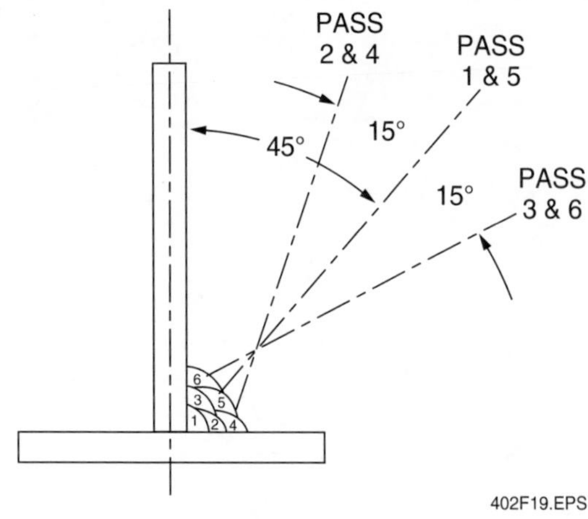

Figure 19 Multiple-pass 2F bead sequence and work angles.

> **NOTE**
>
> Check with your instructor to see if you should run stringer beads or weave beads, or practice both techniques.

5.3.1 Vertical Weave Bead Fillet Welds

Follow these steps to make an uphill fillet weld:

Step 1 Tack two plates together to form a T-joint for the fillet weld coupon.

Step 2 Clamp or tack-weld the coupon in the vertical position.

Step 3 Run the first bead along the root of the joint, starting at the bottom. Use a work angle of about 45 degrees and a 10- to 15-degree push angle (*Figure 20*). Pause in the weld puddle to fill the crater.

Step 4 Clean the weld.

Step 5 Run the remaining passes using a 45-degree work angle and a 10- to 15-degree push angle with a side-to-side weave technique (*Figure 20*). Use a slow motion across the face of the weld. Pause at each toe for penetration and to fill the crater. Clean the weld after each pass.

> **Hot Tip**
>
> **Vertical Fillet Welds**
>
> When making vertical fillet welds, pay close attention to the torch angles. For the first bead, the work angle is approximately 45 degrees. Adjust the work angle for all other welds.

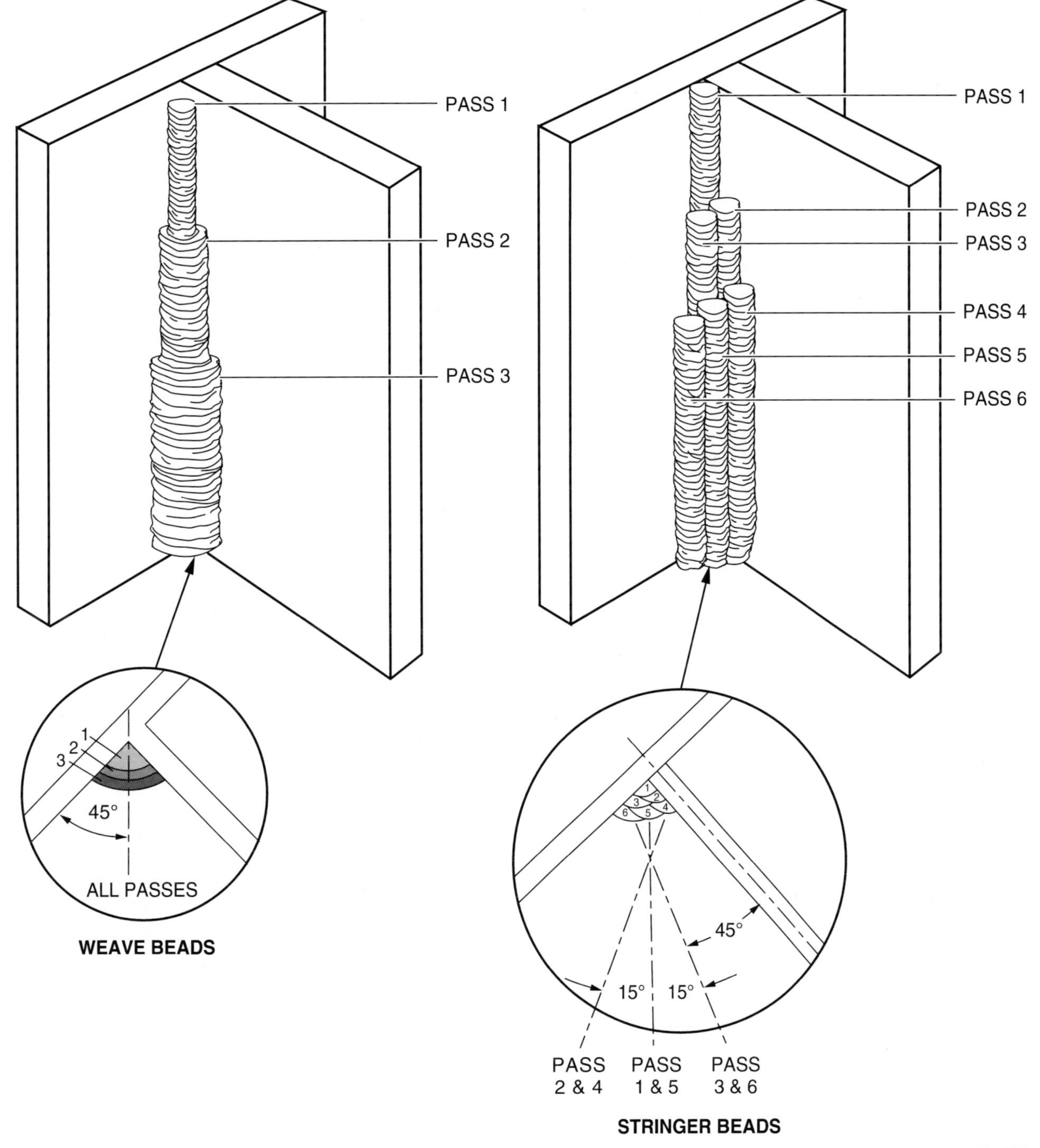

Figure 20 Multiple-pass 3F bead sequences and work angles for stringer and weave beads.

Step 6 Have your instructor inspect the weld. The weld is acceptable if it has the following features:

- Uniform appearance on the bead face
- Craters and restarts filled to the full cross section of the weld
- Uniform weld size of ±1⁄16"
- Acceptable weld profile in accordance with the applicable code or standard
- Smooth transition with complete fusion at the toes of the weld
- No porosity
- No excessive undercut
- No overlap
- No inclusions
- No cracks

5.3.2 *Vertical Stringer Bead Fillet Welds*

Repeat vertical fillet (3F) welding using stringer beads and a slightly oscillating (side-to-side) motion. Pause slightly at each toe to prevent undercut. For stringer beads, use a 10- to 15-degree push angle and the required work angles (*Figure 20*).

5.4.0 Practicing Overhead (4F) Position Fillet Welds

Practice overhead (4F) position fillet welds by making multiple-pass fillet welds in a T-joint. Use an appropriate filler metal as directed by your instructor. When making overhead fillet welds, pay close attention to the torch angles. For the first bead, the work angle is about 45 degrees, and then it is adjusted for all other welds.

Follow these steps to make an overhead fillet weld:

Step 1 Tack two plates together to form a T-joint for the fillet weld coupon.

Step 2 Clamp or tack-weld the coupon so that it is in the overhead position.

Step 3 Run the first bead along the root of the joint. Use a work angle of about 45 degrees and a 10- to 15-degree push angle.

Step 4 Clean the weld.

Step 5 Run the remaining passes at the proper work angles using a 10- to 15-degree push angle and a slight oscillation (*Figure 21*). Overlap each previous pass. Clean the weld after each pass.

Step 6 Have your instructor inspect the weld. The weld is acceptable if it has the following features:

- Uniform appearance on the bead face
- Craters and restarts filled to the full cross section of the weld
- Uniform weld size of ±1/16"
- Acceptable weld profile in accordance with the applicable code or standard
- Smooth transition with complete fusion at the toes of the weld
- No porosity
- No excessive undercut
- No overlap
- No inclusions
- No cracks

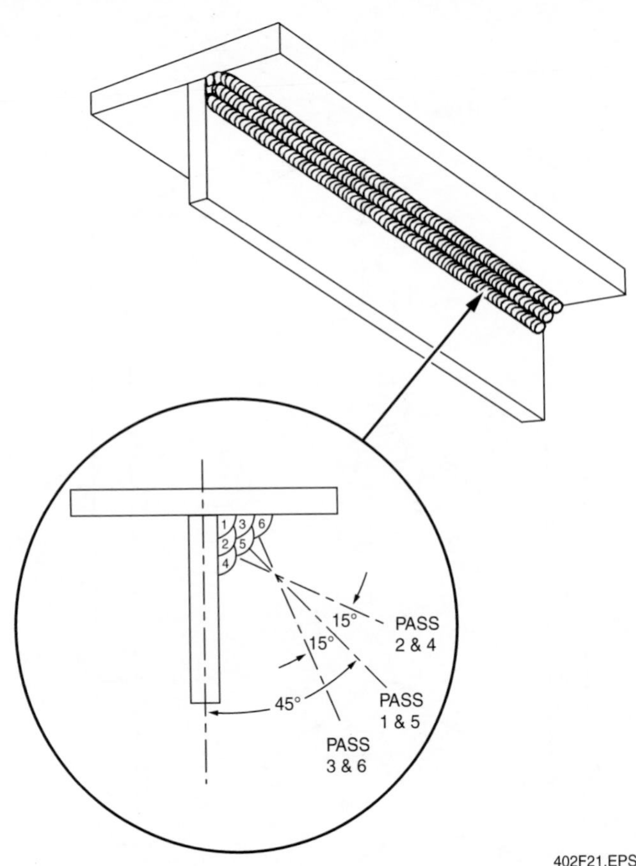

Figure 21 Multiple-pass 4F bead sequence and work angles.

6.0.0 V-Groove Plate Welds

The V-groove weld is a common groove weld made on plate and pipe. The backing method of welding an aluminum V-groove joint with GTAW is presented in this module. Practicing V-groove welds on plate is good preparation for making the more difficult pipe welds, which will be covered later.

6.1.0 Root Pass

For the root pass, use 1/8" aluminum filler metal. After the root pass has been completed, clean and inspect it. Use the appropriate cleaning method to clean the root pass of the oxidation that tends to form along the joint. Inspect the root pass for the following conditions:

- Uniform, smooth face
- Excessive buildup
- Excessive undercut
- Porosity
- Nonfusion

6.2.0 V-Groove Weld Positions

V-groove welds with backing can be made in all positions (*Figure 22*). The weld position for plate is determined by the axis of the weld and the orientation of the workpiece. Groove weld positions are flat (1G), horizontal (2G), vertical (3G), and overhead (4G).

6.3.0 Acceptable and Unacceptable V-Groove Welds with Backing

V-groove welds with backing should be made with slight reinforcement (not exceeding ⅛") and a gradual transition to the base metal at each toe. Groove welds must not have excessive face reinforcement, any underfill, excessive undercut, or any overlap (*Figure 23*). If a groove weld has any of these defects, it must be repaired.

7.0.0 Practicing V-Groove Welds with Backing

Practice making GTAW V-groove welds with backing in the 1G, 2G, 3G, and 4G positions. Use ⅛" aluminum filler metal for the root and other passes. Pay special attention to filling the crater at the termination of the weld. Clean each completed bead with a stainless steel wire brush before starting the next bead.

Hot Tip

Modified U-Groove Joint

Instead of a backing strip, a modified U-groove joint can be used when welding aluminum plate. The modified U-groove joint shown is the same as the one that is used for GTAW of aluminum pipe.

402SA02.EPS

WARNING

Use pliers to handle the hot practice coupons. Wear gloves when placing the practice coupons in water. Steam will rise off the coupons and can burn or scald unprotected hands.

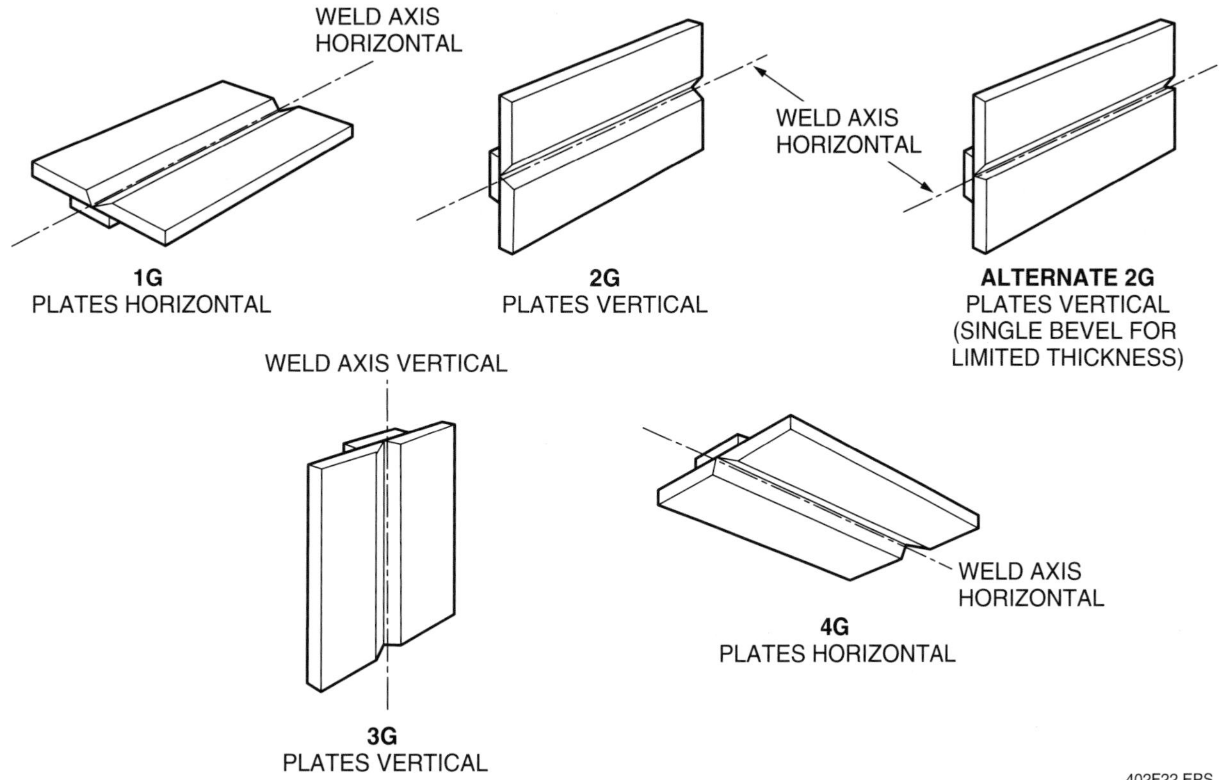

Figure 22 V-groove weld positions.

Hot Tip

Specific V-Groove Requirements

Refer to your site's WPS for specific groove weld requirements. The information in this module is provided as a general guideline only. The WPS or site quality standards must be followed for all welds. Check with your supervisor if you are unsure of the specifications for your application.

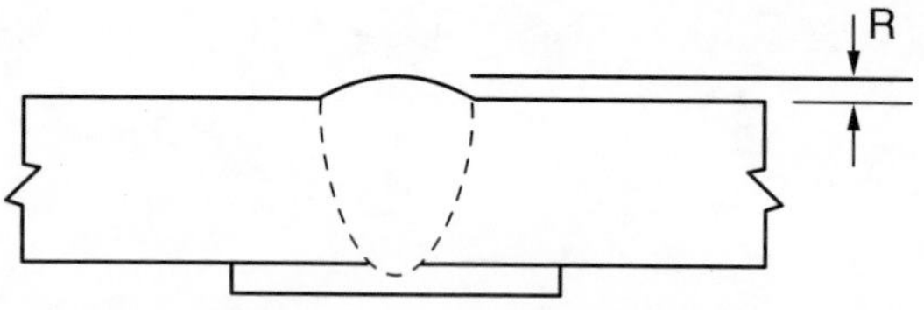

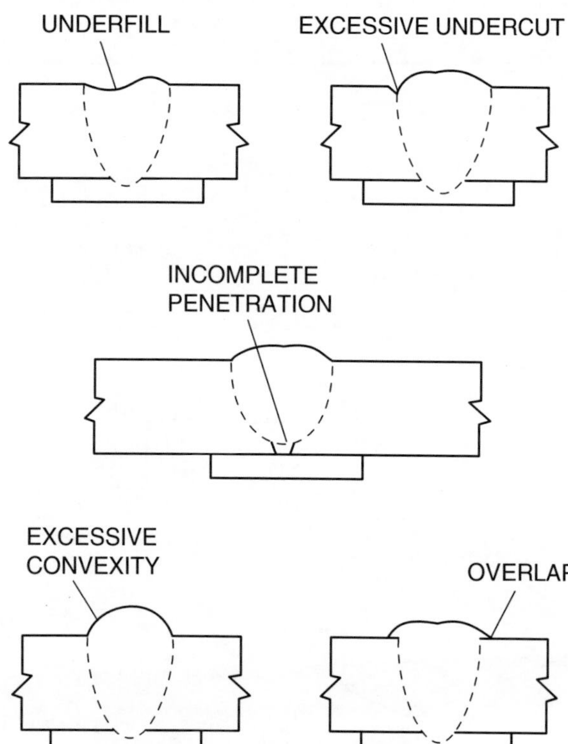

Figure 23 Acceptable and unacceptable V-groove weld profiles.

7.1.0 Practicing Flat (1G) Position V-Groove Welds with Backing

Follow these steps to make V-groove welds with backing in the flat (1G) position (*Figure 24*):

Step 1 Tack-weld the practice coupon together as explained earlier.

Step 2 Clamp or tack-weld the weld coupon in the flat position above the welding table surface.

Step 3 Use a 10- to 15-degree push angle and a 0-degree work angle to run the root pass with an appropriate aluminum filler metal. Feed the filler metal at a 15- to 20-degree angle to the plate.

Step 4 Clean the root pass.

Step 5 Run the remaining passes using a 10-degree to 15-degree push angle and the bead sequence and work angles shown in *Figure 24*. Hold the filler metal about 15 to 20 degrees above the plate face. Clean the weld between each pass.

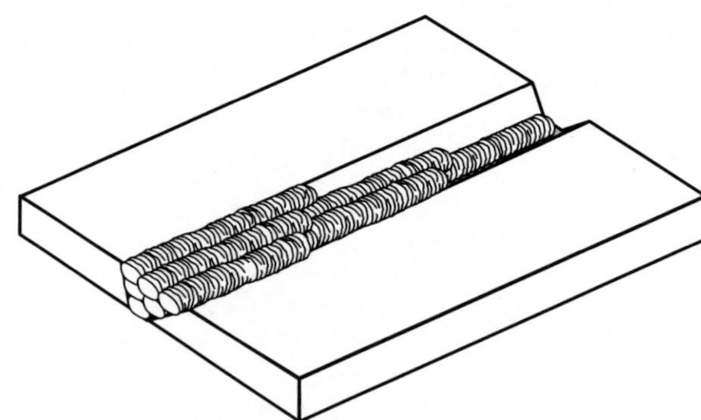

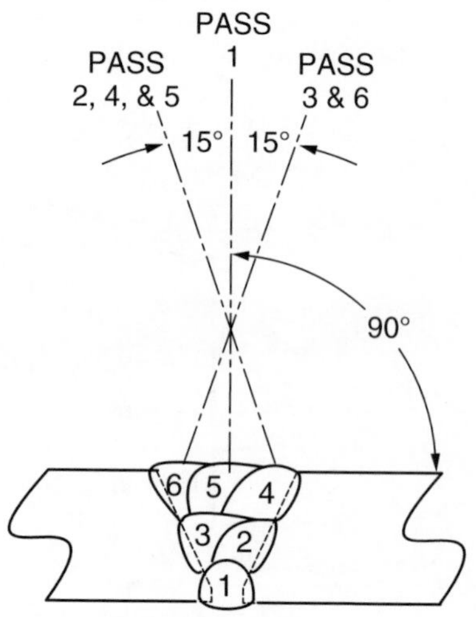

Figure 24 Multiple-pass 1G bead sequence and work angles.

Hot Tip

Cooling Practice Coupons

Cooling coupons with water can affect the mechanical properties of the base metal. Only practice coupons can be cooled with water. Never cool test coupons with water.

7.2.0 Practicing Horizontal (2G) Position V-Groove Welds with Backing

Follow these steps to practice V-groove welds with backing in the horizontal (2G) position (*Figure 25*):

Step 1 Tack-weld the practice coupon together as explained earlier. Use the standard or alternate weld coupon as directed by your instructor.

Step 2 Clamp or tack-weld the coupon in the horizontal position.

Step 3 Run the root pass with an appropriate aluminum filler metal. Use a 10- to 15-degree torch push angle and a 15- to 20-degree filler metal angle with the plate surface. Use the appropriate work angle for Pass 1 shown in *Figure 25*.

Step 4 Clean the weld.

Step 5 Run the remaining passes using a 15- to 20-degree torch push angle and the bead sequence and work angles shown in *Figure 25*. Clean the weld between each pass.

7.3.0 Practicing Vertical (3G) Position V-Groove Welds with Backing

Follow these steps to practice V-groove welds with backing in the vertical (3G) position (*Figure 26*):

Step 1 Tack-weld the practice coupon together as explained earlier.

Step 2 Clamp or tack-weld the coupon in the vertical position.

Step 3 Run the root pass uphill using an appropriate aluminum filler metal. Use a torch push angle of 10 to 15 degrees and a filler metal angle of 15 to 20 degrees with the plate surface. Use a 0-degree work angle for Pass 1 as shown in *Figure 26*.

NOTE: THE ACTUAL NUMBER OF WELD BEADS WILL VARY DEPENDING ON THE PLATE THICKNESS.

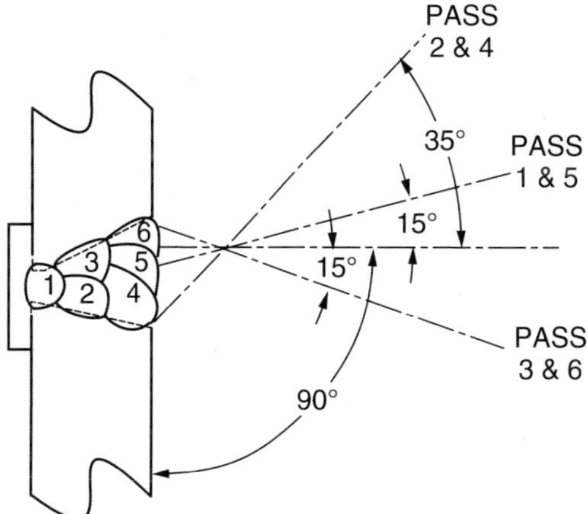

ALTERNATE JOINT REPRESENTATION

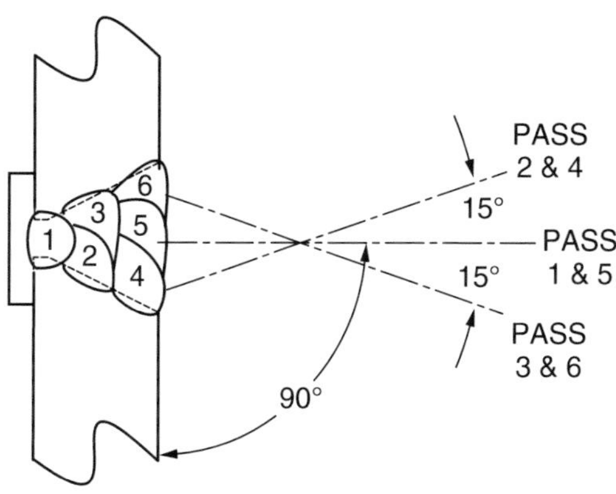

STANDARD JOINT REPRESENTATION

402F25.EPS

Figure 25 Multiple-pass 2G bead sequences and work angles.

Step 4 Clean the weld.

Step 5 Run the remaining passes uphill to complete the weld. Use a 10- to 15-degree torch push angle and the bead sequence and work angles shown in *Figure 26*. Clean the weld between each pass.

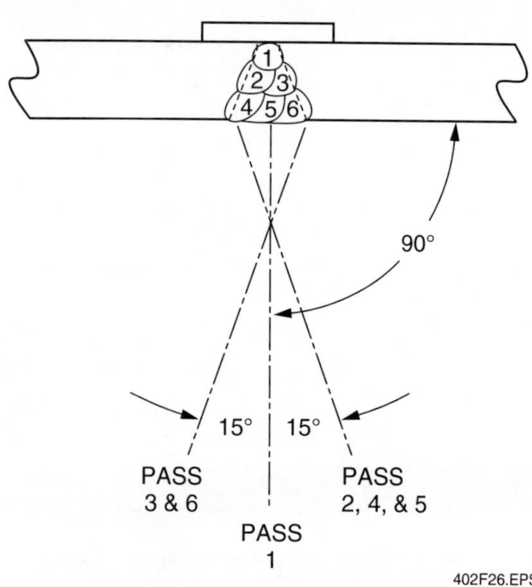

Figure 26 Multiple-pass 3G bead sequence and work angles.

7.4.0 Practicing Overhead (4G) Position V-Groove Welds with Backing

Follow these steps to practice V-groove welds with backing in the overhead (4G) position (*Figure 27*):

Step 1 Tack-weld the practice coupon together as explained earlier.

Step 2 Clamp or tack-weld the coupon in the overhead position.

Step 3 Run the root pass using an appropriate aluminum filler metal. Use a 10- to 15-degree push angle and a 15- to 20-degree filler metal angle with the plate surface. Use a 0-degree work angle for Pass 1, as shown in *Figure 27*.

Step 4 Clean the weld.

Step 5 Run the remaining passes using a 10- to 15-degree torch push angle and the bead sequence and work angles shown in *Figure 27*. Clean the weld between each pass.

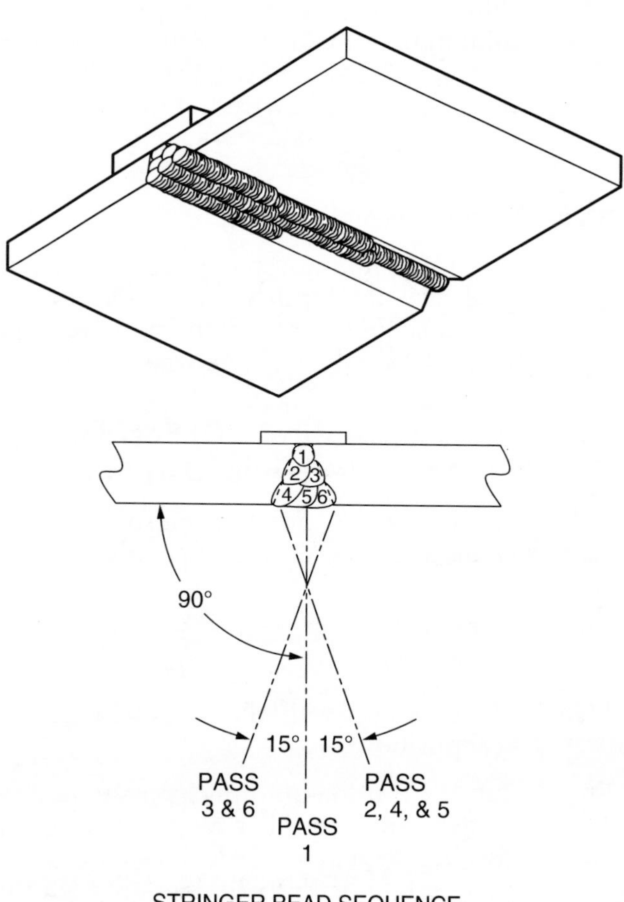

Figure 27 Multiple-pass 4G bead sequence and work angles.

SUMMARY

Welders must understand aluminum metallurgy and be familiar with aluminum's characteristics before they can successfully perform GTAW of aluminum plate. This module covers GTAW equipment and techniques, and it also explains how to select the right filler metals for aluminum welds. In addition, information is provided about how to do the following: make weave and overlapping beads; perform weld terminations and restarts; build a pad on aluminum plate; and make fillet and groove welds on aluminum plate in all positions.

Review Questions

1. Without the use of a properly tinted lens, an electric arc _____.
   a. may safely be viewed indirectly
   b. may safely be viewed directly
   c. may safely be viewed briefly
   d. may never be safely viewed

2. For occasional, very short-term welding of zinc materials, you may use a _____.
   a. large fan
   b. dust mask
   c. full-face supplied-air respirator
   d. HEPA-rated filter on a respirator

3. The aluminum filler metal typically used for the GTAW process is a filler rod that is _____.
   a. dual shield continuous
   b. uncoated solid metal
   c. carbon-coated steel
   d. composite core

4. As a general rule, the aluminum filler rods used for welding aluminum with GTAW are selected to _____.
   a. increase the thermal expansion of the base metal
   b. reduce the thermal expansion of the base metal
   c. add strength to the soft base metal
   d. be compatible with the base metal

5. Filler rods used on aluminum are sometimes cut in half _____.
   a. so their identification numbers can be saved
   b. to make them easier to get into tight locations
   c. so they will fit into the rod holders
   d. to make them easier to handle

6. To clean contamination, such as oil, dirt, and grease, from GTAW filler metal rods used on aluminum, you should use _____.
   a. water
   b. steel wool
   c. compressed air
   d. denatured alcohol

7. When preparing beveled aluminum plate to be welded with a backing strip, how wide must the root opening be?
   a. 1⁄32"
   b. 1⁄16"
   c. 1⁄8"
   d. 3⁄8"

8. The shielding gas most commonly used in GTAW is _____.
   a. argon
   b. helium
   c. hydrogen
   d. carbon dioxide

Review Questions

9. The push angles normally used for GTAW of aluminum plate are _____.
   a. 0 to 10 degrees
   b. 10 to 15 degrees
   c. 20 to 30 degrees
   d. 55 to 65 degrees

10. When using the freehand GTAW technique, how should the tungsten electrode tip be held in relation the weld puddle?
    a. At the trailing edge
    b. Kept in contact with it
    c. Held just above it
    d. Held far above it

11. To ensure proper fusion to the base metal at the weave bead toes, the correct procedure to follow at the edges is to _____.
    a. lift the torch
    b. tilt the torch
    c. proceed more quickly
    d. slow down or pause slightly

12. To make a good weld termination using remote control equipment, as you approach the end of the weld, the welding current should be _____.
    a. immediately turned off
    b. slowly increased
    c. rapidly increased
    d. slowly reduced

13. The weld position in a fillet weld is determined by the orientation of the workpiece and the _____.
    a. size of the electrode
    b. axis of the weld
    c. type of joint
    d. welder's preference

14. In the 2F position, the first bead of the joint should be made with a work angle of approximately _____.
    a. 20 degrees
    b. 45 degrees
    c. 60 degrees
    d. 70 degrees

15. V-groove welds with backing should be made with _____.
    a. the on-the-wire technique
    b. a small amount of overlap
    c. reinforcement not exceeding ⅛"
    d. reinforcement not exceeding ⅜"

Trade Term Introduced in This Module

Helium/argon mixture: A shielding gas mixture in which helium is added to argon to raise the temperature of the arc, which promotes higher welding speeds and deeper weld penetration.

Additional Resources

This module is intended to present thorough resources for task training. The following references are suggested for further study. These are optional materials for continued education rather than for task training.

The Lincoln Electric website offers sources for products and training.
http://www.lincolnelectric.com

The Procedure Handbook of Arc Welding. Cleveland, OH: The James F. Lincoln Arc Welding Foundation, 2000.

Welding Aluminum: Theory and Practice. New York, NY: The Aluminum Association, 2002.

Welding Handbook. Volume 1. *Welding Science & Technology*. Miami, FL: American Welding Society, 2001.

Figure Credits

Lincoln Electric Company, Module opener, 402SA01, 402F10

AWS A5.10/A5.10M:1999 (R2007), Table A2, reproduced with permission of the American Welding Society (AWS), Miami, Florida, 402T01

Terry Lowe, 402F05 (photo), 402F11, 402F12 (photo), 402F14, 402SA02, 402F25 (photo), 402F26 (photo)

The Aluminum Association, Inc., 402T02–402T04

Topaz Publications, Inc., 402F07 (photo), 402F18

Appendix A

Performance Accreditation Tasks

The Performance Accreditation Tasks (PATs) correspond to and support learning objectives in the *AWS EG2.0:2006 Curriculum Guide for the Training of Welding Personnel; Level I – Entry Welder.*

PATs provide specific acceptable criteria for performance and help to ensure a true competency-based welding program for students.

The following tasks are designed to evaluate your ability to run stringer beads, weave beads, overlapping beads, and to make fillet and multiple-pass V-groove welds with GTAW equipment in four standard test positions using aluminum filler wire of the appropriate diameter and shielding gas. Perform each task when you are instructed to do so by your instructor. As you complete each task, show it to your instructor for evaluation. Do not proceed to the next task until told to do so by your instructor. For AWS 2G and 5G certifications, refer to *AWS EG3.0:1996 Guide for the Training and Qualification of Welding Personnel; Level II – Advanced Welder* for bend test requirements. For AWS 6G certifications, refer to *AWS EG4.0:1996 Guide for the Training and Qualification of Welding Personnel; Level III – Expert Welder* for bend test requirements.

WELD A PAD ON ALUMINUM PLATE IN THE FLAT (1G) POSITION USING GTAW STRINGER BEADS

As directed by the instructor, use the GTAW process with the appropriate aluminum filler wire to make the following welds on aluminum plate: stringer beads, weave beads, weld restarts, weld terminations, and overlapping beads.

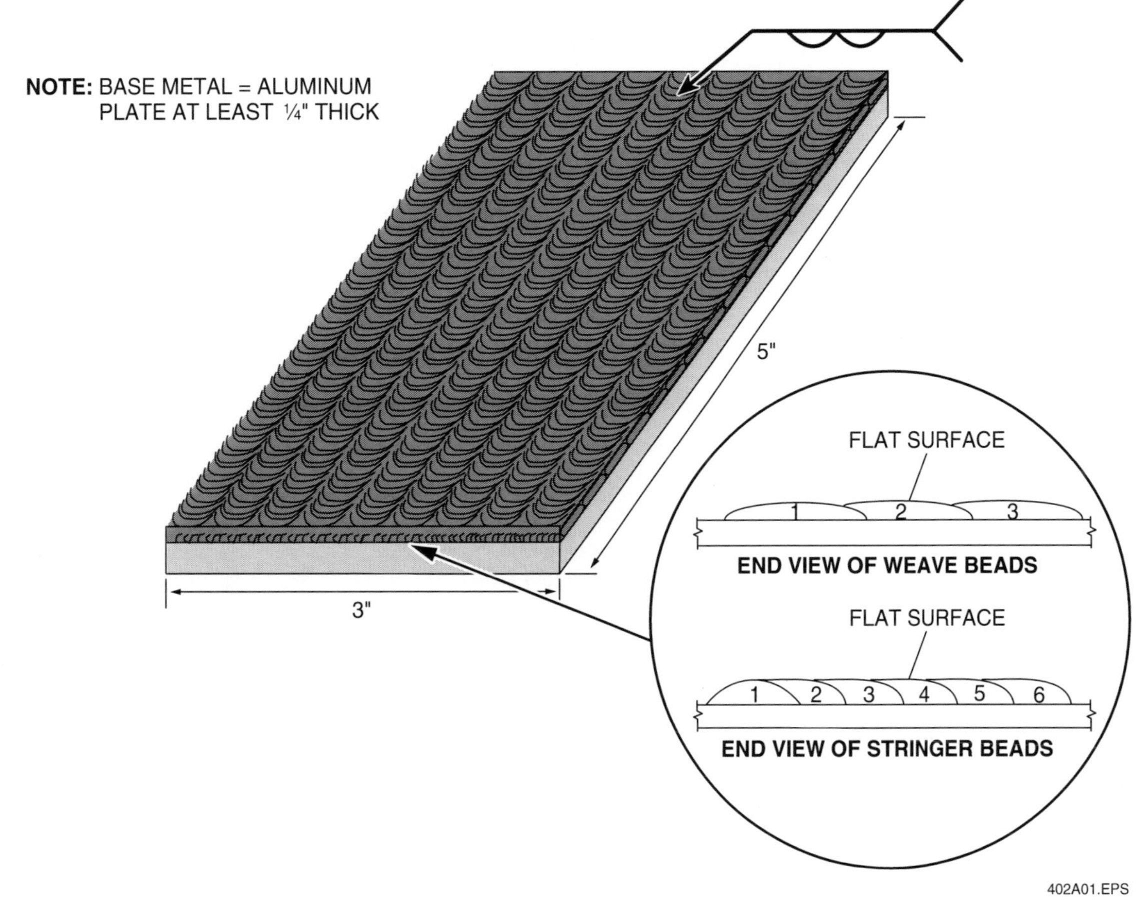

Criteria for Acceptance

- Uniform rippled appearance on the bead face ____________
- Craters and restarts filled to the full cross section of the weld ____________
- Uniform weld width ±1/16" ____________
- Acceptable weld profile in accordance with the acceptable code or standard ____________
- Smooth flat transition with complete fusion at the toes of the weld ____________
- No porosity ____________
- No excessive undercut ____________
- No inclusions ____________
- No cracks ____________

MAKE MULTIPLE-PASS FILLET WELDS ON ALUMINUM PLATE IN THE FLAT (1F) POSITION

As directed by the instructor, use the GTAW process with the appropriate aluminum wire to make a six-pass fillet weld using stringer beads on aluminum plate, as shown.

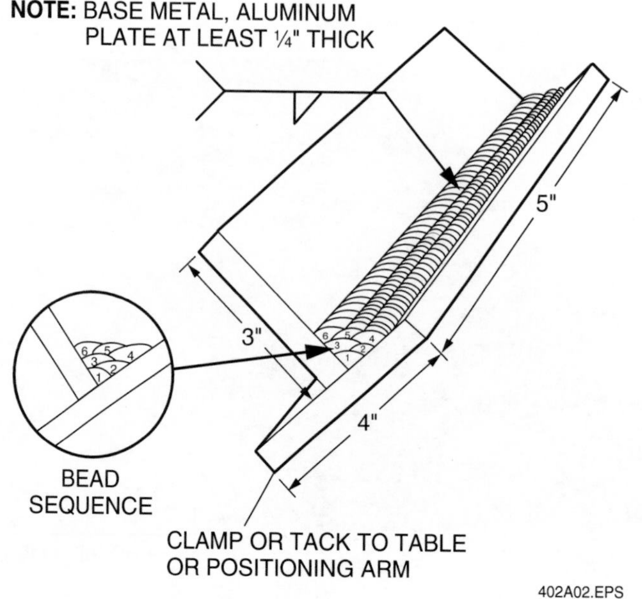

Criteria for Acceptance

- Uniform rippled appearance on the bead face ____________
- Craters and restarts filled to the full cross section of the weld ____________
- Uniform weld size ±1⁄16" ____________
- Smooth flat transition with complete fusion at the toes of the welds ____________
- Acceptable weld profile in accordance with the applicable code or standard ____________
- No porosity ____________
- No excessive undercut ____________
- No overlap ____________
- No inclusions ____________

MAKE MULTIPLE-PASS FILLET WELDS ON ALUMINUM PLATE IN THE HORIZONTAL (2F) POSITION

As directed by the instructor, use the GTAW process with the appropriate aluminum wire to make a six-pass fillet weld using stringer beads on aluminum plate, as shown.

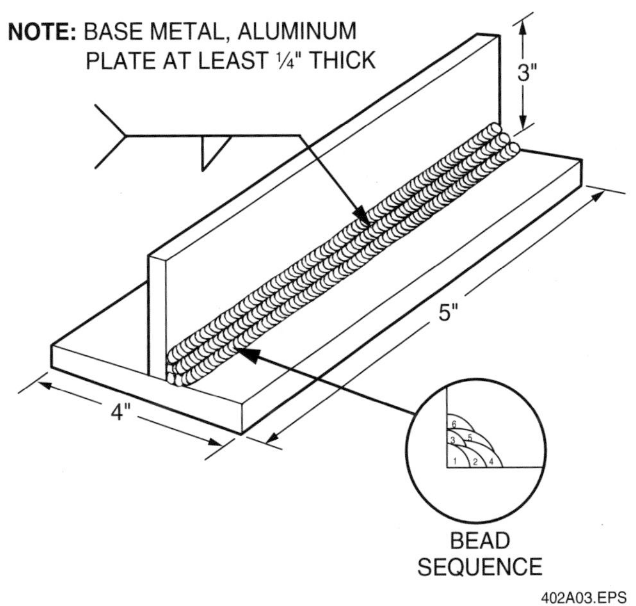

Criteria for Acceptance

- Uniform rippled appearance on the bead face ____________
- Craters and restarts filled to the full cross section of the weld ____________
- Uniform weld size ±1⁄16" ____________
- Smooth flat transition with complete fusion at the toes of the welds ____________
- Acceptable weld profile in accordance with the applicable code or standard ____________
- No porosity ____________
- No excessive undercut ____________
- No overlap ____________
- No inclusions ____________

MAKE MULTIPLE-PASS FILLET WELDS ON ALUMINUM PLATE IN THE VERTICAL (3F) POSITION

As directed by the instructor, use the GTAW process with the appropriate aluminum wire to make a six-pass fillet weld using stringer beads on aluminum plate, as shown.

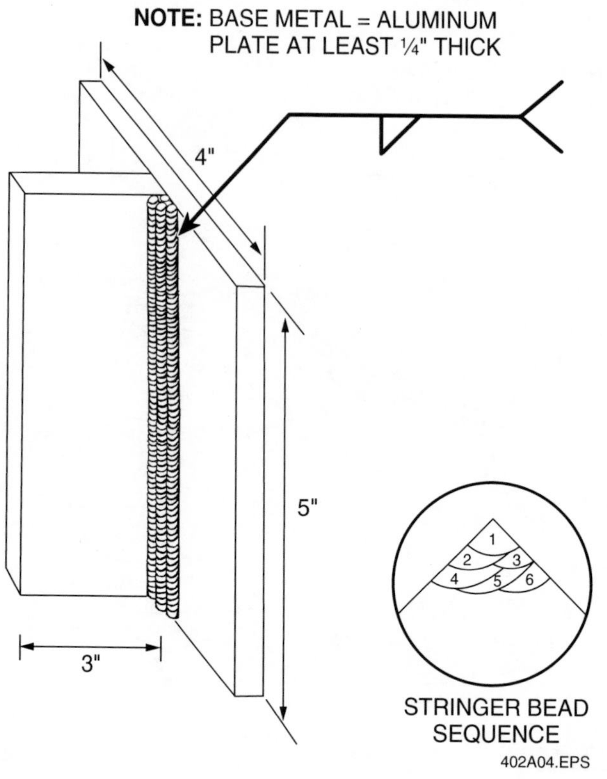

Criteria for Acceptance

- Uniform rippled appearance on the bead face ____________
- Craters and restarts filled to the full cross section of the weld ____________
- Uniform weld width ±1⁄16" ____________
- Acceptable weld profile in accordance with the applicable code or standard ____________
- Smooth flat transition with complete fusion at the toes of the weld ____________
- No porosity ____________
- No excessive undercut ____________
- No inclusions ____________
- No cracks ____________

MAKE MULTIPLE-PASS FILLET WELDS ON ALUMINUM PLATE IN THE OVERHEAD (4F) POSITION

As directed by the instructor, use the GTAW process with the appropriate aluminum wire to make a six-pass fillet weld using stringer beads on aluminum plate, as shown.

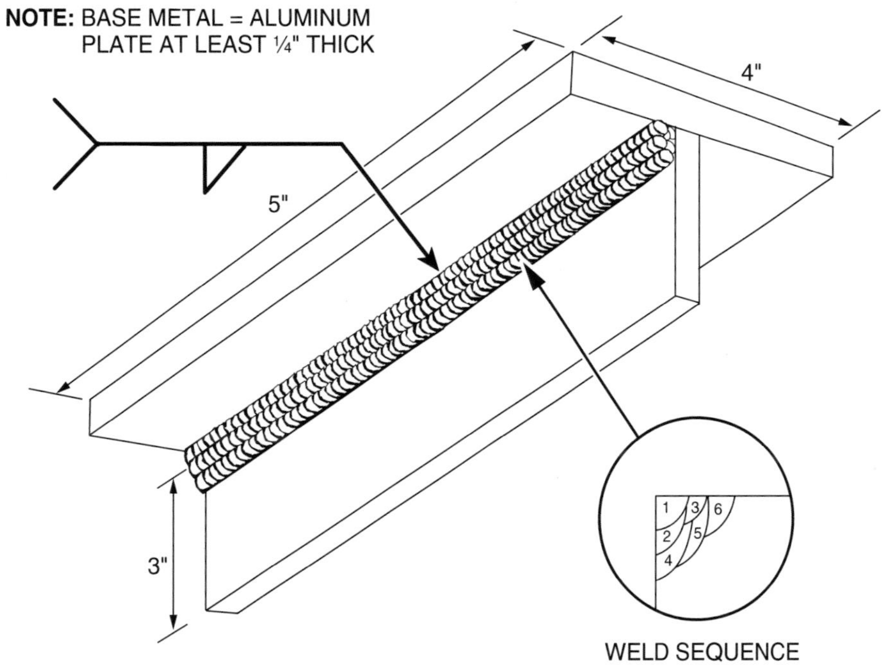

Criteria for Acceptance

- Uniform rippled appearance on the bead face __________
- Craters and restarts filled to the full cross section of the weld __________
- Uniform weld width ±1/16" __________
- Acceptable weld profile in accordance with the applicable code or standard __________
- Smooth flat transition with complete fusion at the toes of the weld __________
- No porosity __________
- No excessive undercut __________
- No inclusions __________
- No cracks __________

MAKE MULTIPLE-PASS V-GROOVE WELDS ON ALUMINUM PLATE WITH BACKING IN THE FLAT (1G) POSITION

As directed by the instructor, use the GTAW process with the appropriate aluminum wire to make a multiple-pass groove weld using stringer beads on aluminum plate with backing, as shown.

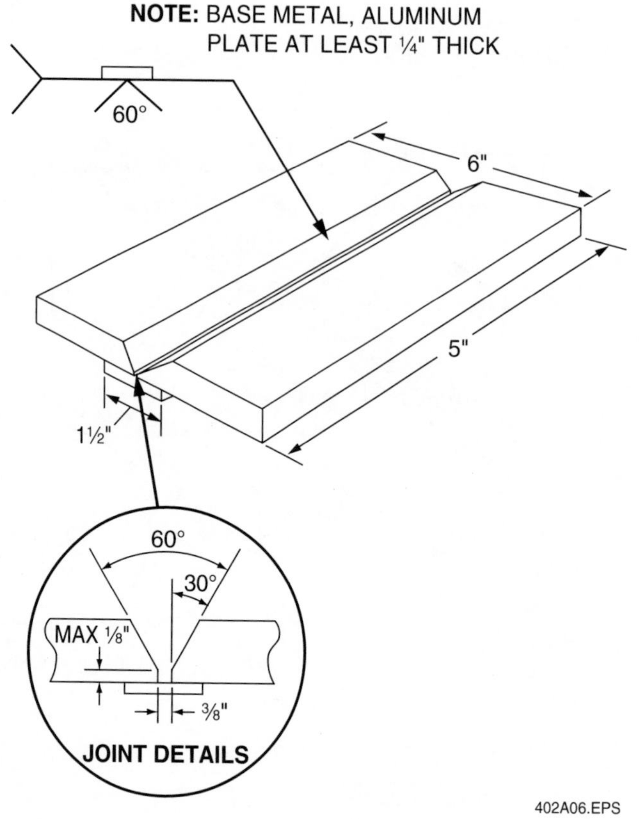

Criteria for Acceptance

- Uniform rippled appearance on the bead face ________
- Craters and restarts filled to the full cross section of the weld ________
- Acceptable weld profile in accordance with the applicable code or standard ________
- Smooth flat transition with complete fusion at the toes of the weld ________
- No porosity ________
- No excessive undercut ________
- No inclusions ________
- No cracks ________

MAKE MULTIPLE-PASS V-GROOVE WELDS ON ALUMINUM PLATE WITH BACKING IN THE HORIZONTAL (2G) POSITION

As directed by the instructor, use the GTAW process with the appropriate aluminum wire to make a multiple-pass groove weld on aluminum plate with backing, as shown.

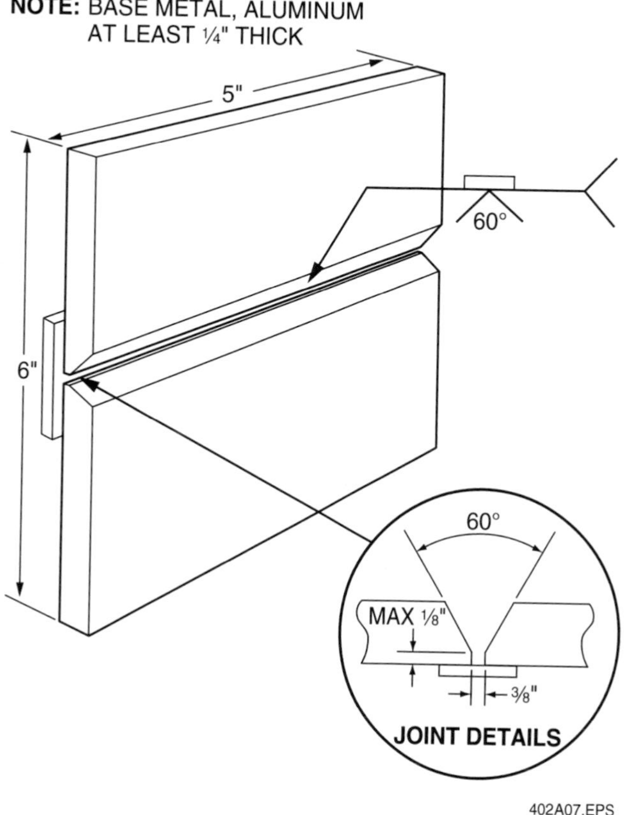

402A07.EPS

Criteria for Acceptance

- Uniform rippled appearance on the bead face ____________
- Craters and restarts filled to the full cross section of the weld ____________
- Acceptable weld profile in accordance with the applicable code or standard ____________
- Smooth flat transition with complete fusion at the toes of the weld ____________
- No porosity ____________
- No excessive undercut ____________
- No inclusions ____________
- No cracks ____________

MAKE MULTIPLE-PASS V-GROOVE WELDS ON ALUMINUM PLATE WITH BACKING IN THE VERTICAL (3G) POSITION

As directed by the instructor, use the GTAW process with the appropriate aluminum wire to make a multiple-pass groove weld on aluminum plate with backing, as shown.

Note: Run the root vertical up.

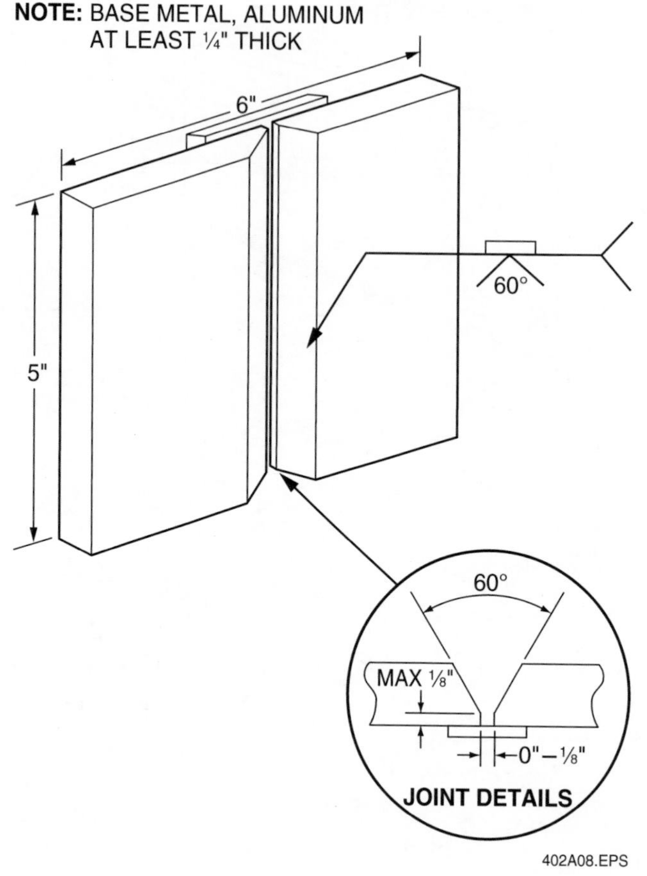

402A08.EPS

Criteria for Acceptance

- Uniform rippled appearance on the bead face ________
- Craters and restarts filled to the full cross section of the weld ________
- Acceptable weld profile in accordance with the applicable code or standard ________
- Smooth flat transition with complete fusion at the toes of the weld ________
- No porosity ________
- No excessive undercut ________
- No inclusions ________
- No cracks ________

MAKE MULTIPLE-PASS V-GROOVE WELDS ON ALUMINUM PLATE WITH BACKING IN THE OVERHEAD (4G) POSITION

As directed by the instructor, use the GTAW process with the appropriate aluminum wire to make a multiple-pass groove weld on aluminum plate with backing, as shown.

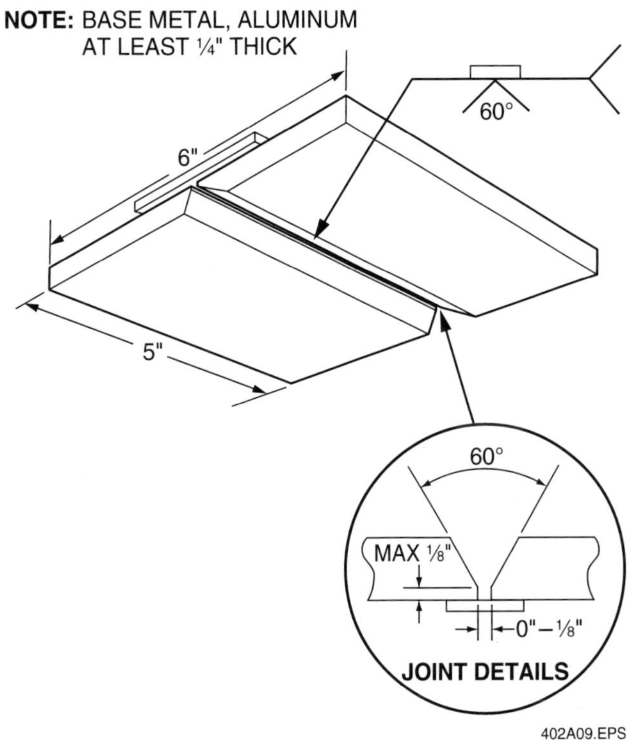

Criteria for Acceptance

- Uniform rippled appearance on the bead face ____________
- Craters and restarts filled to the full cross section of the weld ____________
- Acceptable weld profile in accordance with the applicable code or standard ____________
- Smooth flat transition with complete fusion at the toes of the weld ____________
- No porosity ____________
- No excessive undercut ____________
- No inclusions ____________
- No cracks ____________

NCCER CURRICULA — USER UPDATE

NCCER makes every effort to keep its textbooks up-to-date and free of technical errors. We appreciate your help in this process. If you find an error, a typographical mistake, or an inaccuracy in NCCER's curricula, please fill out this form (or a photocopy), or complete the online form at **www.nccer.org/olf**. Be sure to include the exact module ID number, page number, a detailed description, and your recommended correction. Your input will be brought to the attention of the Authoring Team. Thank you for your assistance.

Instructors – If you have an idea for improving this textbook, or have found that additional materials were necessary to teach this module effectively, please let us know so that we may present your suggestions to the Authoring Team.

NCCER Product Development and Revision

13614 Progress Blvd., Alachua, FL 32615

Email: curriculum@nccer.org
Online: www.nccer.org/olf

❑ Trainee Guide ❑ AIG ❑ Exam ❑ PowerPoints Other ______________________________

Craft / Level: Copyright Date:

Module ID Number / Title:

Section Number(s):

Description:

Recommended Correction:

Your Name:

Address:

Email: Phone:

WELDING

The sport of car racing relies heavily on the craft of welding. With proper welds, a car can reach top speeds while keeping the driver safe.

FRAM

Close-up view of roll bar welds.

Gas tungsten arc welding (GTAW or TIG) and ga metal arc welding (GMAW or MIG) are two wel processes often used in motorsports.

GMAW (MIG) welding of roll bars.

GTAW (TIG) welding of header pipes.

Racing teams and NASCAR require precision welding of roll bars. Careful work here can mean the difference between life and death for the driver.

GTAW – Aluminum Pipe

29403-10

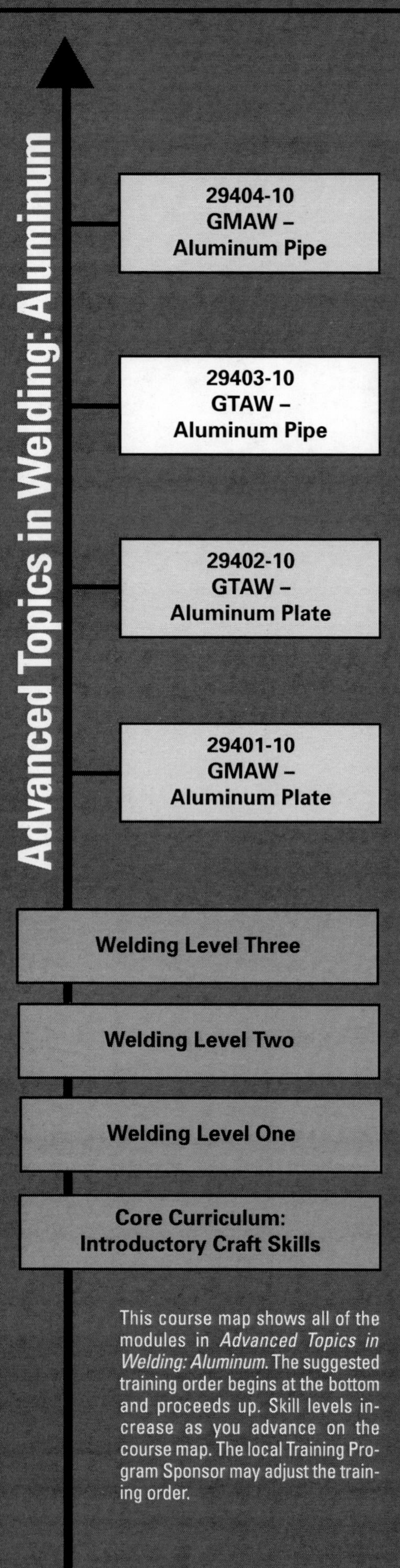

This course map shows all of the modules in *Advanced Topics in Welding: Aluminum.* The suggested training order begins at the bottom and proceeds up. Skill levels increase as you advance on the course map. The local Training Program Sponsor may adjust the training order.

Objectives

When you have completed this module, you will be able to do the following:

1. Prepare GTAW equipment for V-groove and modified U-groove welds on aluminum pipe.
2. Identify and explain V-groove and modified U-groove welds on aluminum pipe with GTAW equipment.
3. Perform V-groove and modified U-groove welds on aluminum pipe in the following positions using GTAW equipment:
   - 2G
   - 5G
   - 6G

Prerequisites

Before you begin this module, it is recommended that you successfully complete *Core Curriculum*; *Welding Level One*; *Welding Level Two*; *Welding Level Three*; and *Advanced Topics in Welding: Aluminum*, Modules 29401-10 and 29402-10.

Contents

Topics to be presented in this module include:

Figures and Tables

1.0.0 Introduction

Gas tungsten arc welding (GTAW) uses an arc between a tungsten electrode and the base metal to melt the base and filler metals. The electrode, the arc, and the molten base metal are shielded from atmospheric contamination by a flow of inert gas from the torch nozzle (*Figure 1*). The filler metal is a rod whose composition is compatible with the base metal. The filler metal is usually handheld and manually fed into the leading edge of the weld puddle. GTAW produces high-quality welds without slag or oxidation. Because there is no flux, there is no corrosion due to slag entrapment. As a result, minimal postweld cleaning is needed.

A major use of GTAW is to make manual high-quality welds on aluminum piping that is used in both critical and noncritical applications. The GTAW process allows greater control of root penetration and fill than almost any other process. For this reason, GTAW is often used to make the root pass on pipe, even when the remaining passes are made with GMAW, which has a higher deposition rate.

This module covers how to set up GTAW equipment. It also explains how to perform V-groove or modified U-groove welds on aluminum pipe with aluminum filler metal in the 2G, 5G, and 6G welding positions. The dimensions and specifications in this module are representative of codes in general and may not be specific to any particular code. Always follow the proper codes for your site.

2.0.0 Safety Summary

The following is a summary of safety procedures and practices that must be observed when welding. Keep in mind that this is only a summary; complete safety coverage is provided in the Level One module, *Welding Safety*. If you have not completed that module, do so before continuing.

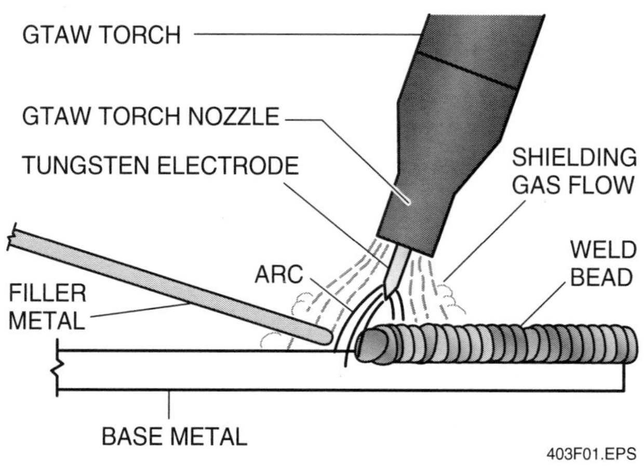

Figure 1 GTAW process.

Above all, always use the proper protective clothing and equipment when welding.

2.1.0 Protective Clothing and Equipment

Welding work creates flying debris, such as sparks or small chunks of hot metal. Anyone welding or assisting a welder must use the proper protective clothing and equipment. The following list provides protective clothing and equipment guidelines:

- Always use safety glasses with a full-face shield or a helmet. The glasses, face shield, or helmet lens must have the proper light-reducing tint for the type of welding being performed. Never directly or indirectly view an electric arc without using a properly tinted lens.
- Wear the proper protective leather and/or flame retardant clothing and welding gloves (*Figure 2*). They will protect you from flying sparks, molten metal, and heat.
- Wear 8-inch or taller high-top safety shoes or boots. Make sure that the tongue and lace area is covered by a pant leg. Sometimes the tongue and lace area is exposed or the footwear must be protected from burn marks. In those cases, wear leather spats under the pants or chaps and over the front and top of the footwear.
- Wear a solid material (non-mesh) hat with a bill pointing to the back or toward the ear closest to the welding. This will give added protection. If much overhead welding is required, use a full leather hood with a welding faceplate and the correct tinted lens. If a hard hat is required, use one that allows the attachment of both rear deflector material and a face shield.
- Wear earmuffs or earplugs to protect ear canals from sparks.

Figure 2 Welding while wearing protective equipment.

2.2.0 Fire/Explosion Prevention

Welding work includes the cutting, grinding, and welding of metal. All of these actions generate heat; and often produce flying sparks. The heat and flying sparks can be the cause of fires and explosions. Welders must use extreme care to protect both themselves and others near their work. The following list provides fire and explosion protection guidelines:

- Never carry matches or gas-filled lighters in your pockets. Sparks can cause the matches to ignite or the lighter to explode, causing serious injury.
- Never perform any type of heating, cutting, or welding until a hot work permit has been obtained and an approved fire watch established. Most work-site fires in these types of operations are started by cutting torches.
- Never use oxygen to blow dust or dirt from clothing. The oxygen can remain trapped in the fabric for a time. If a spark hits the clothing during this period, the clothing can burn rapidly and violently out of control.
- Make sure that any flammable material in the work area is either moved or shielded by a fire-resistant covering. Approved fire extinguishers must be available before attempting any heating, welding, or cutting operations.
- Always comply with any site requirement for a hot work permit or fire watch.
- Never release a large amount of fuel gas, especially acetylene. Methane and propane tend to concentrate in and along low areas. Both gases can ignite at a considerable distance from the release point. Acetylene is lighter than air, but it is even more dangerous than methane. When mixed with air or oxygen, acetylene will explode at much lower concentrations than any other fuel.
- To prevent fires, maintain a neat and clean work area. Also make sure that any metal scrap or slag is cold before disposal.

GOING GREEN

Water Disposal

To protect the environment and save resources, make sure to properly dispose of any water used in the cutting of tanks, barrels, or various metals. If the water can be reused, save it and use it again for the next cutting.

WARNING

Before welding containers, such as tanks or barrels, always check to see if they have ever held any explosive, hazardous, or flammable materials. These include, but are not limited to, petroleum products, citrus products, or chemicals that decompose into toxic fumes when heated. As standard practice, always clean and then fill any tanks or barrels with water, or purge them with an appropriate purging gas to displace any oxygen.

2.3.0 Work Area Ventilation

Welders normally work within inches of their welds wearing special protective helmets. Vapors from the welds can be hazardous. The following list provides work area ventilation guidelines:

- Always follow the required confined space procedures before conducting any welding in the confined space.
- Never use oxygen to ventilate confined spaces.

WARNING

An oxygen monitor may be required when working in a confined space.

- Always perform welding operations in a well-ventilated area (*Figure 3*). Welding operations involving zinc or cadmium materials or coatings result in toxic fumes. For long-term welding of these materials, always wear an approved, full-face SAR that uses breathing air supplied from outside of the work area. For occasional, very short-term exposure, you may use a HEPA-rated or metal-fume filter on a standard respirator.
- Make sure that confined spaces are properly ventilated for welding operations.

3.0.0 Welding Preparation

Before welding can begin, the area must be readied, the welding equipment set up, and the metal to be welded prepared. The following sections explain how to set up the equipment for welding.

To practice welding, you will need a welding table, bench, or stand. The welding surface can be steel, but an aluminum surface is preferred. Provisions must be available for placing weld coupons out of position.

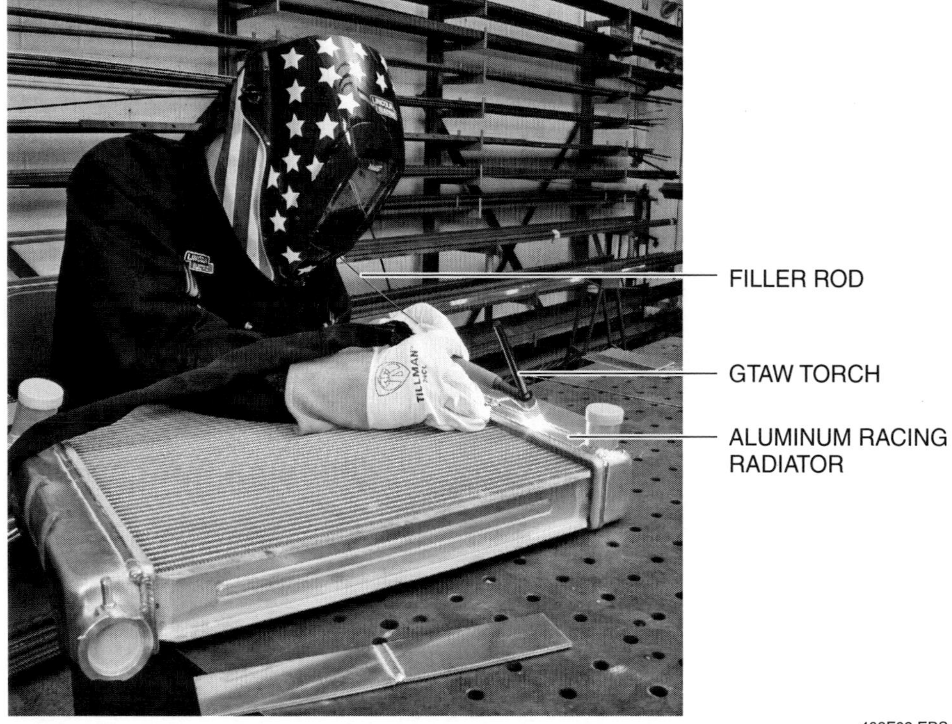

Figure 3 Welding on aluminum in a well-ventilated area.

To set up the area for welding, follow these steps:

Step 1 Check to be sure that the area is properly ventilated. Make use of doors, windows, and fans.

Step 2 Check the area for fire hazards. Remove any flammable materials before proceeding.

Step 3 Locate the nearest fire extinguisher. Do not proceed unless the extinguisher is charged and you know how to use it.

Step 4 Set up flash shields around the welding area.

GOING GREEN

Conserve Materials

Pipe for practice welding is expensive and difficult to obtain. To conserve resources and reduce waste in landfills, completely use weld coupons until all surfaces have been welded upon. Cutting coupons apart and reusing the pieces conserves materials. Check with your instructor for the proper size of coupon.

3.1.0 Practice Pipe Weld Coupons

Pipe weld coupons should be cut from Schedule 40 aluminum pipe that is 3" to 12" in diameter. Each welded joint requires two coupons of the same size. *Figure 4* shows typical AWS aluminum pipe specifications for bevel angles, modified U-grooves, root faces, and root openings. For this module, the practice coupons are tacked together with no root openings or backing rings. The coupons are also used without any backing gas.

To prepare weld coupons for groove weld joints, follow these steps:

Step 1 Use an approved solvent to remove any grease or oil within a minimum of 1" from the weld joint. Clean the inside and outside of the aluminum pipe with a grinder or stainless steel brush.

WARNING

If you use a grinding wheel, make sure it is a special high-speed grinding wheel approved for use on aluminum. It should not be used on any other type of metal or on surfaces contaminated with grease or oil. Using other types of grinding wheels on aluminum may cause the grinding wheels to shatter.

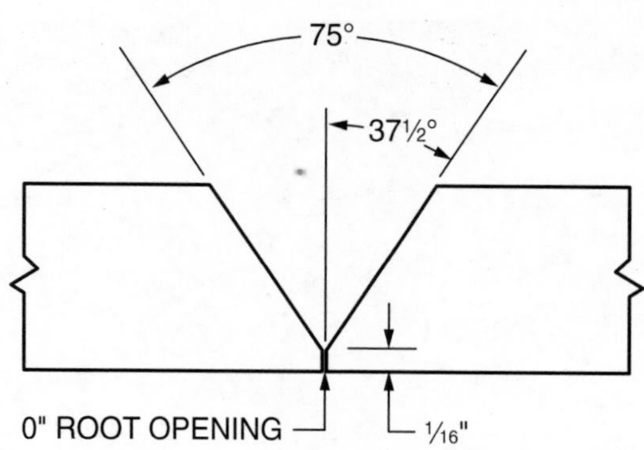

STANDARD V-GROOVE
(WALL THICKNESS ¾" OR LESS)

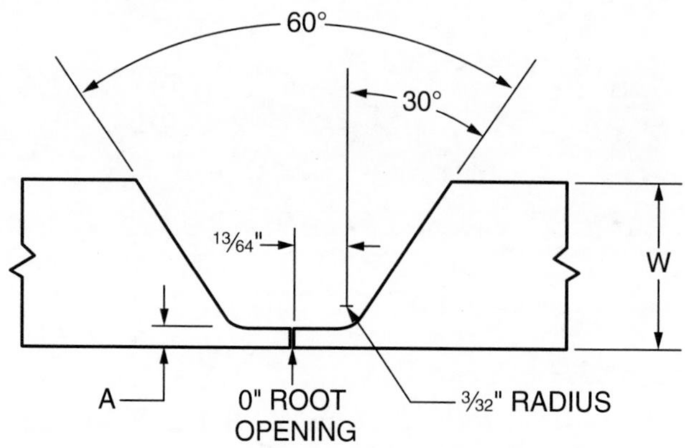

MODIFIED U-GROOVE (PREFERRED) FOR AC GTAW

PIPING DIMENSIONS				
NOMINAL PIPE DIAMETER SIZE NUMBER		OUTSIDE DIAMETER (OD): EXTENDED ROOT FACE THICKNESS		WALL THICKNESS
DN	NPS		A	W (Max)
6 – 65	⅛ – 2½"	13/32 – 2⅞"	1/16"	9/32"
80 – 300	3 – 12"	3½ – 12¾"	3/32"	½"

403F04.EPS

Figure 4 AWS aluminum pipe groove specifications for use without backing.

CAUTION

To prevent contamination, do not use a stainless steel wire brush on anything else after it has been used on aluminum.

WARNING

To prevent injury, always comply with the MSDS guidelines for the cleaning solvents being used.

Step 2 Bevel or groove the end of the pipe, as shown in *Figure 4*, using any acceptable method. Techniques include mechanical cutting, thermal cutting, or grinding. The preferred GTAW joint for aluminum pipe without backing is the modified U-groove.

Step 3 Cut off a section of the beveled pipe end (4" minimum).

Step 4 Check the groove. There should be no dross. The groove angle should be as shown in *Figure 4*, with no notches deeper than 1/16".

Hot Tip

Machined Grooves and Root Faces

Machining the grooves and root faces of aluminum pipe is the cleanest and most uniform way to prepare pipe for GTAW. This is especially true for aluminum when all contaminants from thermal cutting must be removed to eliminate weld discontinuities. Most aluminum pipe is available with the ends premachined for the desired groove.

Step 5 Grind or file a root face on the bevel as specified by your instructor or as shown in *Figure 4*.

NOTE

If the groove bevel has been cut thermally, machine the face ⅛" back from the face of the bevel.

Step 6 Align the two pipe sections so that their ID surfaces are even all around. Align small-diameter pipe by clamping both pieces to a piece of angle iron. Align large-diameter pipe with the aid of a pipe alignment jig or by holding a straightedge across the joint, parallel to the pipe axis. The straightedge must be used all around the pipe in case one or both sections are distorted.

Step 7 When the pipe ends are aligned with no root opening, make the first of four tack welds.

NOTE

Heavy-wall or large-diameter pipe may require longer tacks or more than four tacks.

Step 8 Make the second tack weld on the opposite side from the first, making sure there is no root opening.

Hot Tip

Welding Large-Diameter Pipe

The practice coupons on large-diameter pipe should be greater than 6" in length. They need more metal to help absorb the increased heat used to make these welds.

Step 9 Check the root again, and weld the third tack midway between the first two tacks.

Step 10 Weld the fourth tack opposite the third tack and midway between the first and second tacks. There should now be four tack welds evenly spaced (90 degrees) around the pipe coupon (*Figure 5*).

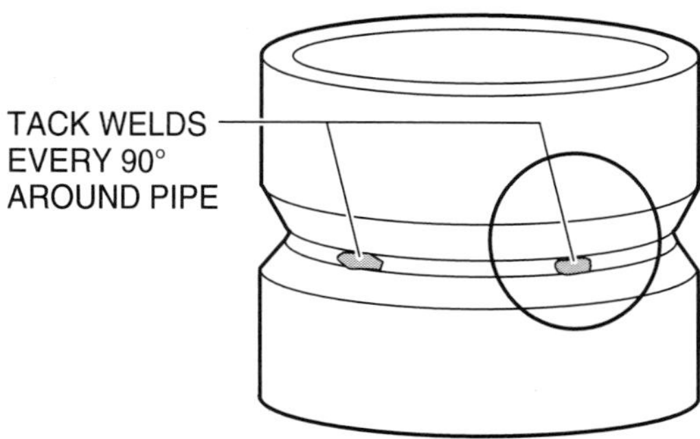

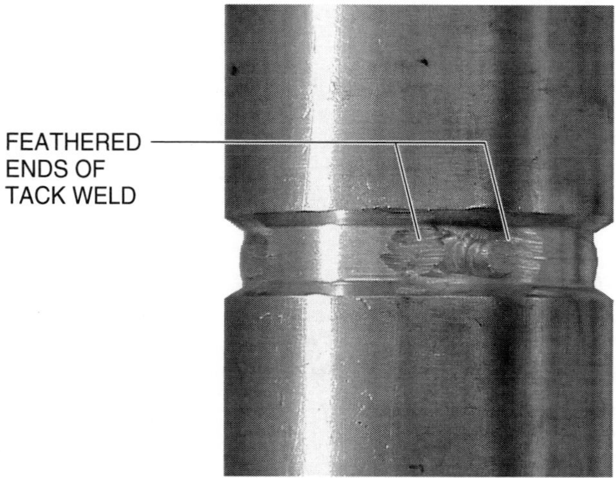

403F05.EPS

Figure 5 Tacked U-groove weld coupon.

Hot Tip

Aluminum Pipe Weld Backing

Open-root and consumable insert welding do not generally work well with aluminum piping. When you need to make an open-root groove weld, you must back it with a temporary or permanent backing ring; otherwise, you must eliminate the root opening when you use a standard V-groove or modified U-groove joint configuration. No gas backing is required when either of the above methods is used. The modified U-groove joint is recommended for manual AC GTAW.

Step 11 Feather the tack welds with the edge of a grinding wheel. Feathering the ends of the tack welds with a grinder helps to fuse the root pass into the tack weld.

3.2.0 The Welding Machine

Identify a proper constant-current welding machine for GTAW, and follow these steps to set it up for use:

Step 1 Verify that the welding machine can be used for GTAW, with or without internally controlled gas shielding.

Step 2 Identify an air-cooled or water-cooled GTAW torch. Make sure that it is compatible with the welding machine and any cooling unit.

Step 3 Verify the location of the primary disconnect.

Step 4 Configure the welding machine for GTAW parameters as directed by your instructor (*Figure 6*). Configure the torch polarity to AC. Equip the torch with a properly prepared $\frac{3}{32}$" or $\frac{1}{8}$" EWP (pure) or EWZr (zirconiated) tungsten electrode.

> **NOTE**
>
> Pure tungsten (EWP) electrodes are used for welding aluminum and magnesium with AC and high frequency. Zirconiated tungsten (EWZr) electrodes are used for welding aluminum and magnesium with AC and high frequency for welds in which tungsten inclusions are not tolerated and higher current capacity is desired.

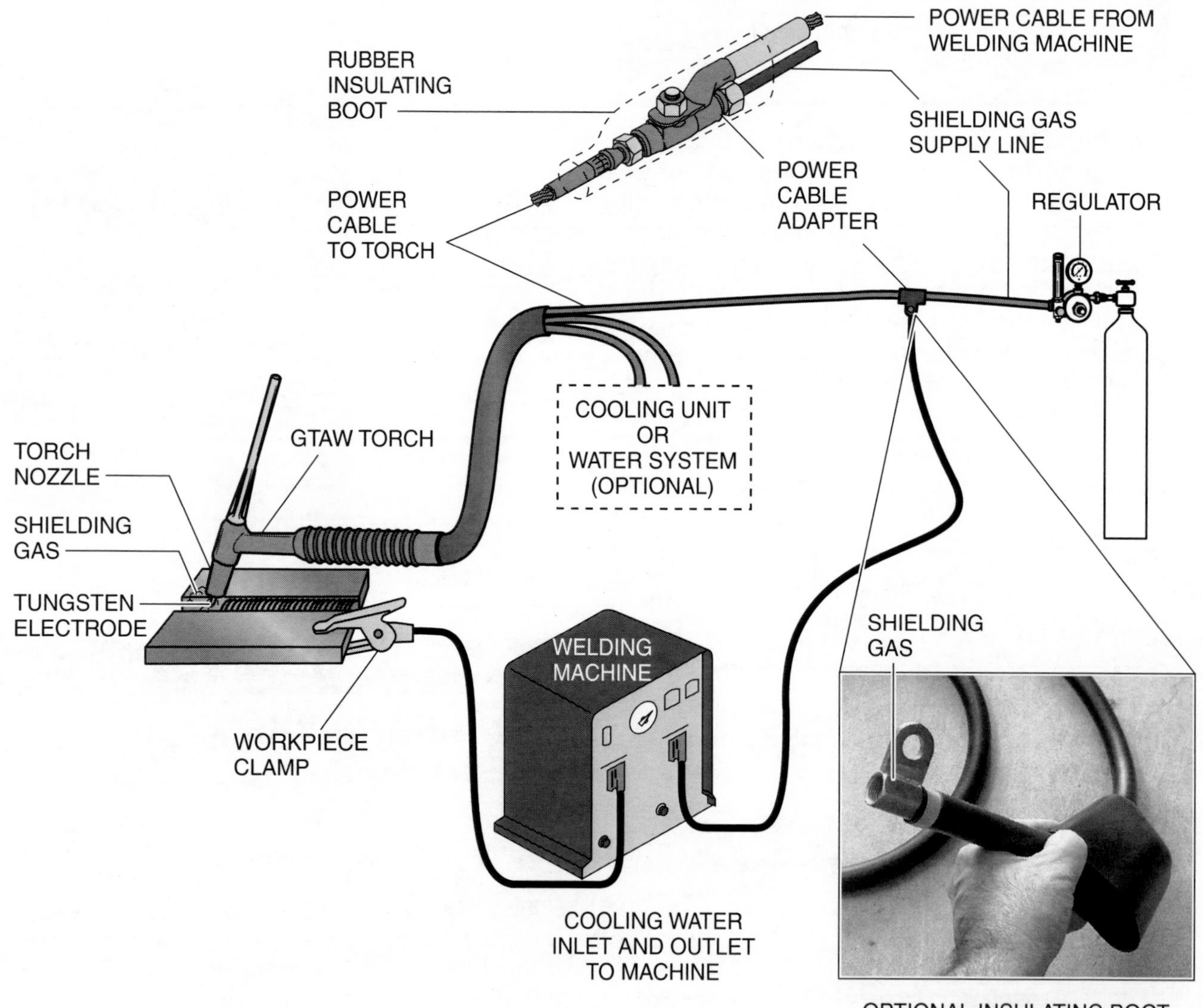

Figure 6 Configuration diagram of a typical GTAW machine.

Step 5 Connect the proper shielding gas for the application as described in the previous level and as specified by the filler metal manufacturer, WPS, site quality standards, or your instructor.

Step 6 Connect the clamp of the workpiece lead to the workpiece.

Step 7 Turn on the welding machine, and purge the torch as directed by the manufacturer's instructions.

3.3.0 Filler Metals

Filler metals are selected to be compatible with the base metal to be welded. The WPS or site standards will specify the proper filler metal type and size to use.

> **CAUTION**
>
> When the WPS or site quality standards specify a filler metal, it must be used to prevent weld discontinuities.

For the welding exercises in this module, weld the aluminum pipe coupons using 3⁄32" and/or 1⁄8" aluminum filler metals, as specified by your instructor. Remove only a small number of filler rods at a time. Keep the remainder in the package to keep them clean. Before using the filler metal, check it for burned ends or contamination, such as corrosion, dirt, oil, or grease, which can all cause weld discontinuities. Chemically clean the filler metal with a clean rag. Use emery cloth or stainless steel wool to remove oxides, and snip any burned ends. If the filler metal cannot be cleaned, do not use it.

4.0.0 Gas Tungsten Arc Welding Techniques

GTAW weld bead characteristics and quality are affected by several factors, each of which is influenced by the way the welder handles the torch. These factors include the following:

- Torch travel speed and arc length
- Torch angles
- Torch and filler metal handling techniques

GOING GREEN

Advanced Inverter Power Sources

An advanced inverter power source can make welding aluminum with GTAW easier, while saving electrical energy. With features such as current pulsing, weld current sequencing (stop/start ramping), and AC frequency adjustment, these machines provide much more control over some of the variables that affect the quality of a weld.

4.1.0 Torch Travel Speed and Arc Length

Torch travel speed and arc length affect the GTAW weld puddle and weld penetration. Slower travel speeds allow more heat to concentrate and form a larger, more deeply penetrating puddle. Faster travel speeds prevent heat buildup and form smaller, shallower puddles. Arc length is the major control for bead width. As the torch is raised, voltage, arc length, and bead width increase.

4.2.0 Torch Angles

The two basic torch angles that must be controlled when performing GTAW are the work angle and the travel angle. The definition of these angles is the same as for all other methods of arc welding.

4.2.1 Work Angle

The torch work angle (*Figure 7*) is a less-than-90-degree angle between a line perpendicular to the major workpiece surface at the point of electrode contact and a plane determined by the electrode axis and the weld axis. For a T-joint or corner joint, the line is perpendicular to the nonbutting member. For pipe, the plane is determined by the electrode axis and a line tangent to the pipe surface at the same point.

4.2.2 Travel Angle

The torch travel angle (*Figure 7*) is a less-than-90-degree angle between the electrode axis and a line perpendicular to the weld axis at the point of

> **Hot Tip**
>
> **Advanced Excessive Push Angle**
>
> A push angle that is too large tends to draw air from under the back edge of the torch nozzle where the air mixes with the shielding gas stream and contaminates the weld.

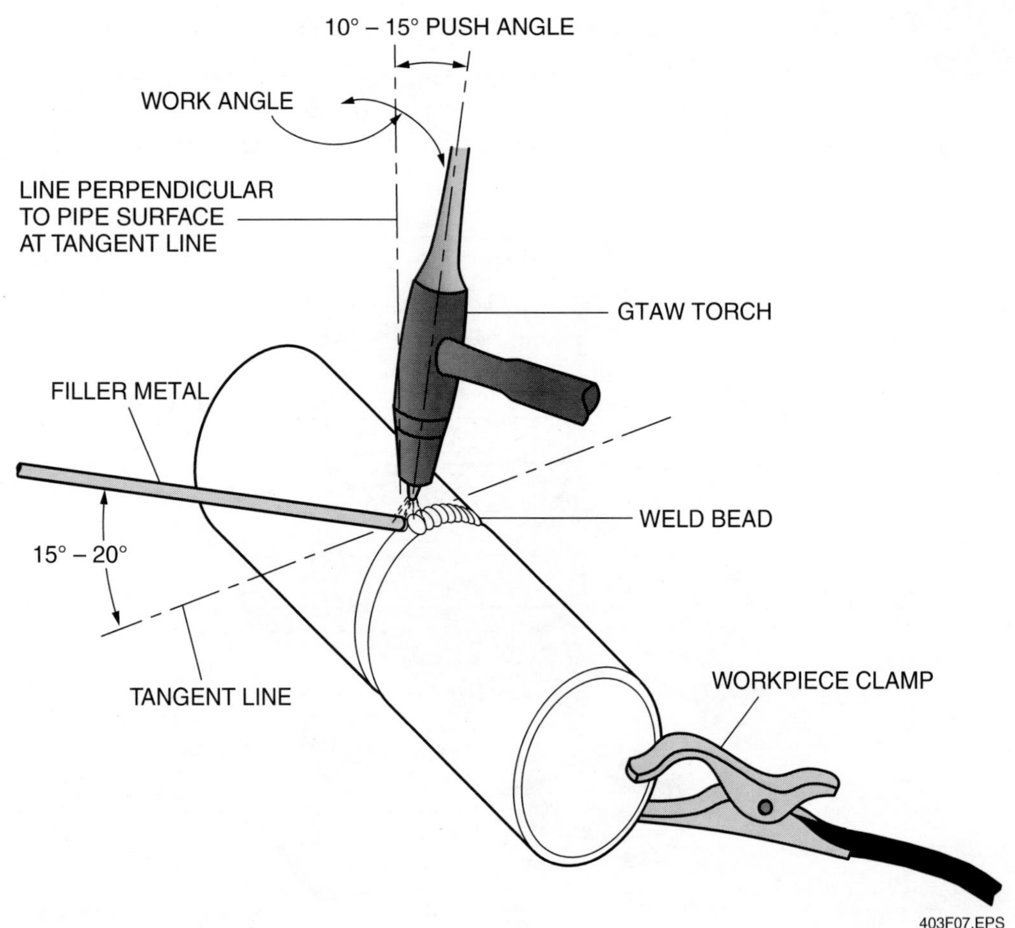

Figure 7 Typical torch work and travel angles.

electrode contact in a plane determined by the electrode axis and the weld axis. For pipe, the plane is determined by the electrode axis and a line tangent to the pipe's surface at the same point. A push angle is used for GTAW. A push angle is created when the torch is tilted back so that the electrode tip precedes the torch in the direction of the weld. In this position, the electrode tip and shielding gas are directed ahead of the weld bead. This provides base metal cleaning on the positive portion of the AC voltage cycle. Push angles of 10 to 15 degrees are normally used for GTAW of aluminum.

4.3.0 Torch and Filler Metal Handling Techniques

The two basic torch and filler metal handling methods used to perform GTAW are the freehand and the walking-the-cup techniques, which were covered in previous modules. Try both and use the one that gives the best results.

5.0.0 V-Groove and Modified U-Groove Pipe Welds

The modified U-groove and the V-groove welds are typically used for joining medium- and thick-walled pipe used in critical piping systems. Critical piping systems include pipe, fittings, and welded joints. These systems contain or carry material that may cause long- or short-term catastrophic danger or damage to people and/or the environment. The damage can occur when the piping system fails to contain or carry the material as designed. Welds in critical piping must meet the most strict code requirements. Noncritical piping is low-pressure piping used for heating and air conditioning, simple water systems, and other service installations. The code requirements used to evaluate noncritical piping welds are less strict than those used to assess welds in critical piping systems. Although modified U-groove and V-groove welds made with backing are the most common welds performed on aluminum pipe, no backing will be used for the exercises in this module.

CAUTION

When welding pipe, avoid making arc strikes outside of the weld groove on the surface of the pipe. An arc strike can cause a hardened spot that can crack as the pipe expands and contracts. An arc strike on the pipe surface is considered a defect, which will require repair or reworking.

Hot Tip

Weld Pool Size

When performing GTAW in the horizontal, vertical, and overhead positions, you must keep the weld pool smaller than it is in the flat position. The weld pool is generally held in the joint by surface tension. When the pool is too large, gravitational force overcomes the capillary action, and the liquid metal runs out of the joint.

5.1.0 Techniques for Aluminum Pipe GTAW

Aluminum pipe is commonly welded with alternating current using argon as a shielding gas. In some cases, an argon/helium mix is used for increased penetration. Higher currents are used when welding aluminum because of high thermal conductivity in the base metal. The modified U-groove is preferred for joints without backing because it decreases the heat conducted away from the weld joint and allows complete root penetration and fusion with a smaller weld pool. This smaller weld pool is more easily controlled in both vertical and overhead positions by surface tension and molecular attraction. Without backing material, the most difficult part of welding is getting the proper penetration and fusion for the root pass. The challenge is to establish a weld pool that penetrates and fuses with the pipe without falling through.

5.2.0 Pipe Groove Weld Test Positions

Groove welds may be made in all positions on pipe. The weld position is determined by the axis of the pipe. Four standard weld test positions are used with pipe (*Figure 8*).

To determine the quality of a pipe weld, specific sections of pipe are removed for destructive testing. For positions 1G-ROTATED and 2G, two specimens are removed. They must be taken from opposite sides of the pipe, as shown in *Figure 9*. Once removed, the specimens are put through different bend tests to determine the quality of the welds.

To destructively test 5G or 6G welds, the test specimens are cut from the four regions shown in *Figure 9*. These four regions are midway between the 12-o'clock and 3-o'clock, 3-o'clock and 6-o'clock, 6-o'clock and 9-o'clock, and 9-o'clock and 12-o'clock positions.

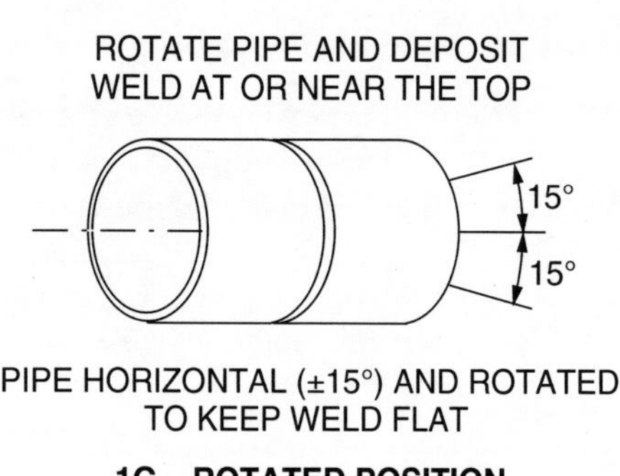

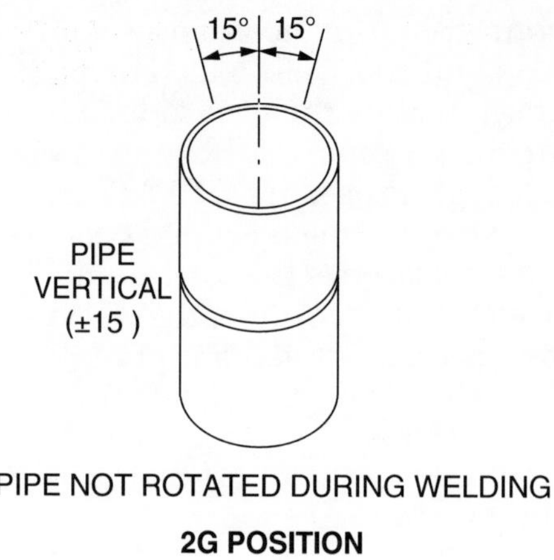

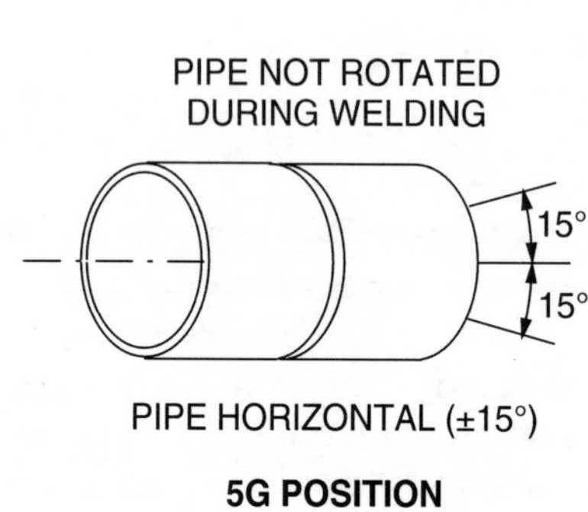

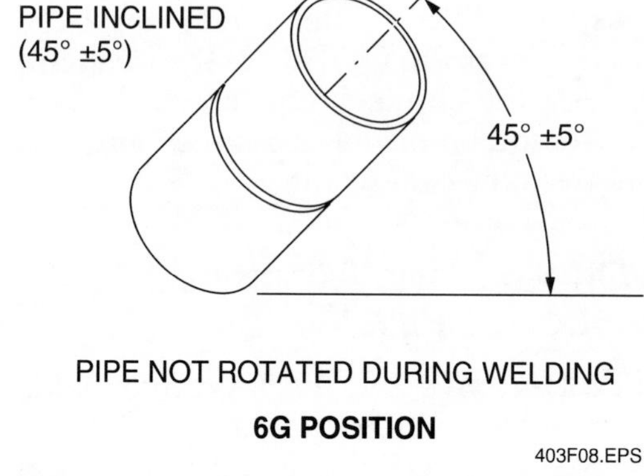

Figure 8 Four basic pipe groove weld test positions.

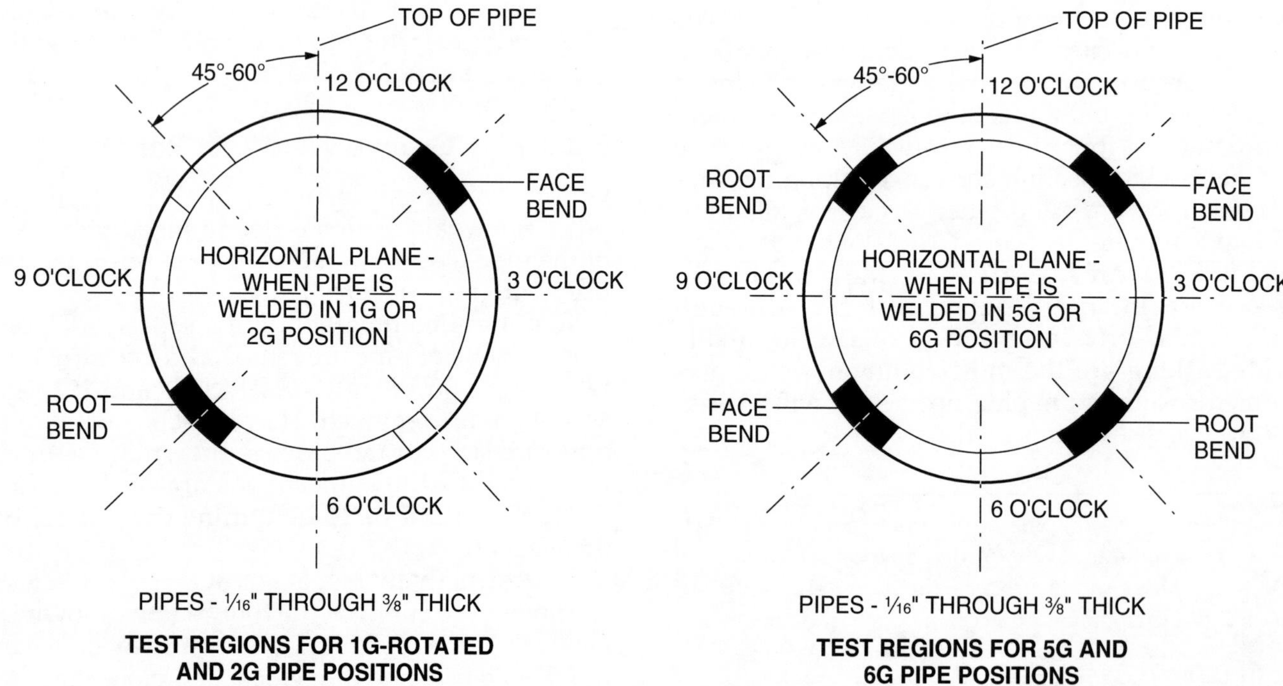

Figure 9 Test specimen regions for pipe positions 1G-ROTATED, 2G, 5G, and 6G.

Hot Tips

Test Position

Welding pipe in the 6G position is a common welding test. A ring may be added to test for restricted accessibility (6GR).

Filler Metal

Clean the filler metal rods with an approved chemical to remove hydrocarbon contaminants, and then clean them with stainless steel wool to remove any oxides.

5.3.0 Acceptable and Unacceptable Pipe Weld Profiles

Pipe groove welds without backing should be made with slight reinforcement (not to exceed ⅛") and a gradual transition to the base metal at each toe. The root pass should have complete penetration. The root reinforcement on the inside of the pipe ranges from being flush to a maximum of ⅛". Pipe groove welds must not have excessive reinforcement or any underfill at the face or root. They must be free from excessive undercut and any overlap. Excessively large face reinforcement reduces the pipe's strength. It causes the stresses in the pipe to be concentrated along the sides of the weld. As a result, the pipe will not expand and contract uniformly along its length. If a weld has any of these defects (*Figure 10*), it must be repaired.

6.0.0 Practicing V-Groove or Modified U-Groove Pipe Welds

The following sections explain how to perform these V-groove or modified U-groove weld positions:

- Horizontal (2G)
- Multiple (5G)
- Multiple inclined (6G)

6.1.0 Practicing Horizontal (2G) Position Groove Welds

Practice 2G position groove pipe welds. Use argon shielding gas and a filler rod that has a diameter specified by your instructor. The torch should be at a 10- to 15-degree push angle for the weld passes (*Figure 11*). To prevent the weld puddle from sagging, the torch work angle can be dropped slightly, but not more than 10 degrees. For the root pass, use a circular motion to form the weld puddle. Be sure to fuse into the tack welds.

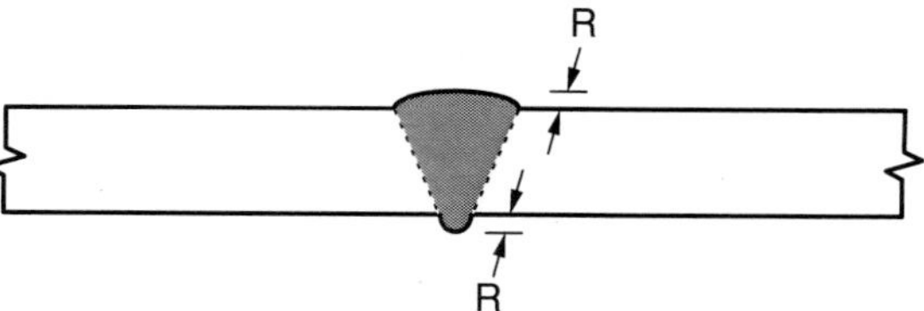

R = FACE AND ROOT REINFORCEMENT PER CODE NOT TO EXCEED ⅛" MAX.

ACCEPTABLE WELD PROFILE

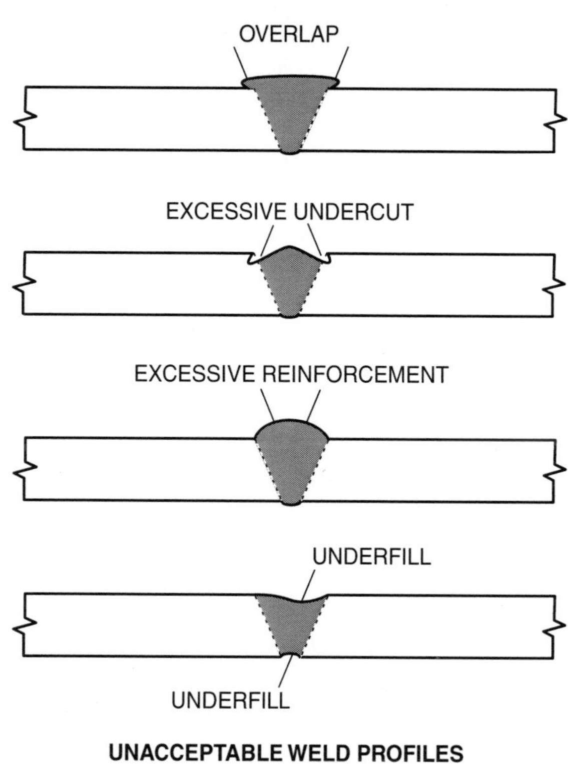

UNACCEPTABLE WELD PROFILES

403F10.EPS

Figure 10 Acceptable and unacceptable pipe groove weld profiles.

When running the remaining passes, use stringer beads and either the walking-the-cup or the freehand torch handling technique. Ensure proper fusion at the toes of the weld bead. Take special care to fill the crater at the termination of the weld.

Follow these steps to practice groove pipe welds in the 2G position:

Step 1 Tack-weld together the practice pipe weld coupon as explained earlier.

Step 2 Clamp or tack-weld the pipe coupon with the axis of the pipe vertical.

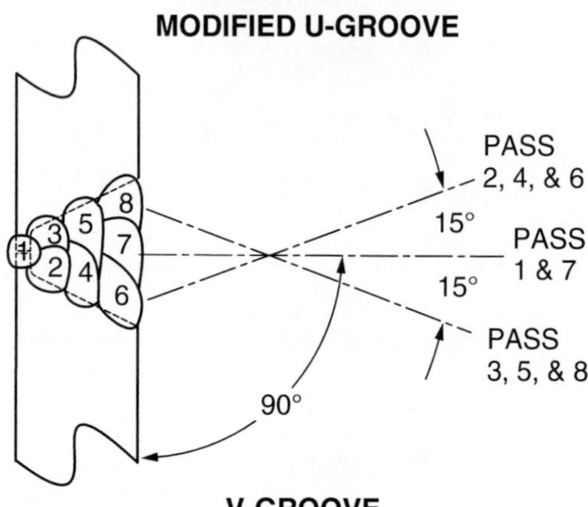

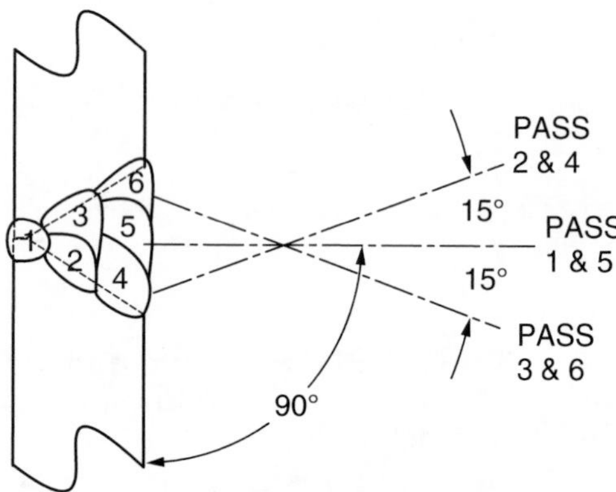

Figure 11 Multiple-pass 2G bead sequences and work angles.

Step 3 Run the root pass (Pass 1) as shown in *Figure 11*, starting on a tack weld. Continue this procedure until the root pass has been completed.

Step 4 Brush the root pass to clean the weld.

Step 5 Run the remaining passes at the proper work angles. Overlap the passes, and clean the weld after each pass.

6.2.0 Practicing Multiple (5G) Position Groove Welds

Practice the multiple (5G) position groove pipe welds. Use argon shielding gas and a solid filler rod that has a diameter specified by your instructor. Start on a tack weld positioned at the bottom of the pipe, and weld uphill toward the top. Use a torch push angle of 10 to 15 degrees to the pipe surface (*Figure 12*).

The torch should be at approximately a 10- to 15-degree push angle for the root and remaining passes (*Figure 13*). When running the remaining passes, use stringer beads and either the walking-the-cup or the freehand torch handling technique. Ensure proper fusion at the toes of the weld bead. Take special care to fill the crater at the termination of the weld.

Follow these steps to practice groove pipe welds in the 5G position:

Step 1 Tack-weld together the practice pipe weld coupon as explained earlier.

Step 2 Clamp or tack-weld the pipe weld coupon into position with the pipe axis horizontal.

Step 3 Run the root pass uphill with a stringer bead, as shown in *Figure 12*. Repeat the procedure for the opposite side of the pipe.

Step 4 Brush the root pass to clean the weld.

Step 5 Run the remaining passes uphill at the proper work angles. Overlap the passes, and clean the weld after each pass.

6.3.0 Practicing Multiple Inclined (6G) Position Groove Welds

Practice the 6G multiple inclined (45-degree) position groove pipe welds. Use a shielding gas and a solid filler rod that has a diameter specified by your instructor. Start at the bottom of the pipe, and weld uphill toward the top.

The torch should be at approximately a 10- to 15-degree push angle for the root and remaining passes (*Figure 14*). When running the remaining passes, use stringer beads and either the walking-the-cup or the freehand torch handling technique. Ensure proper fusion at the toes of the weld bead. Take special care to fill the crater at the termination of the weld.

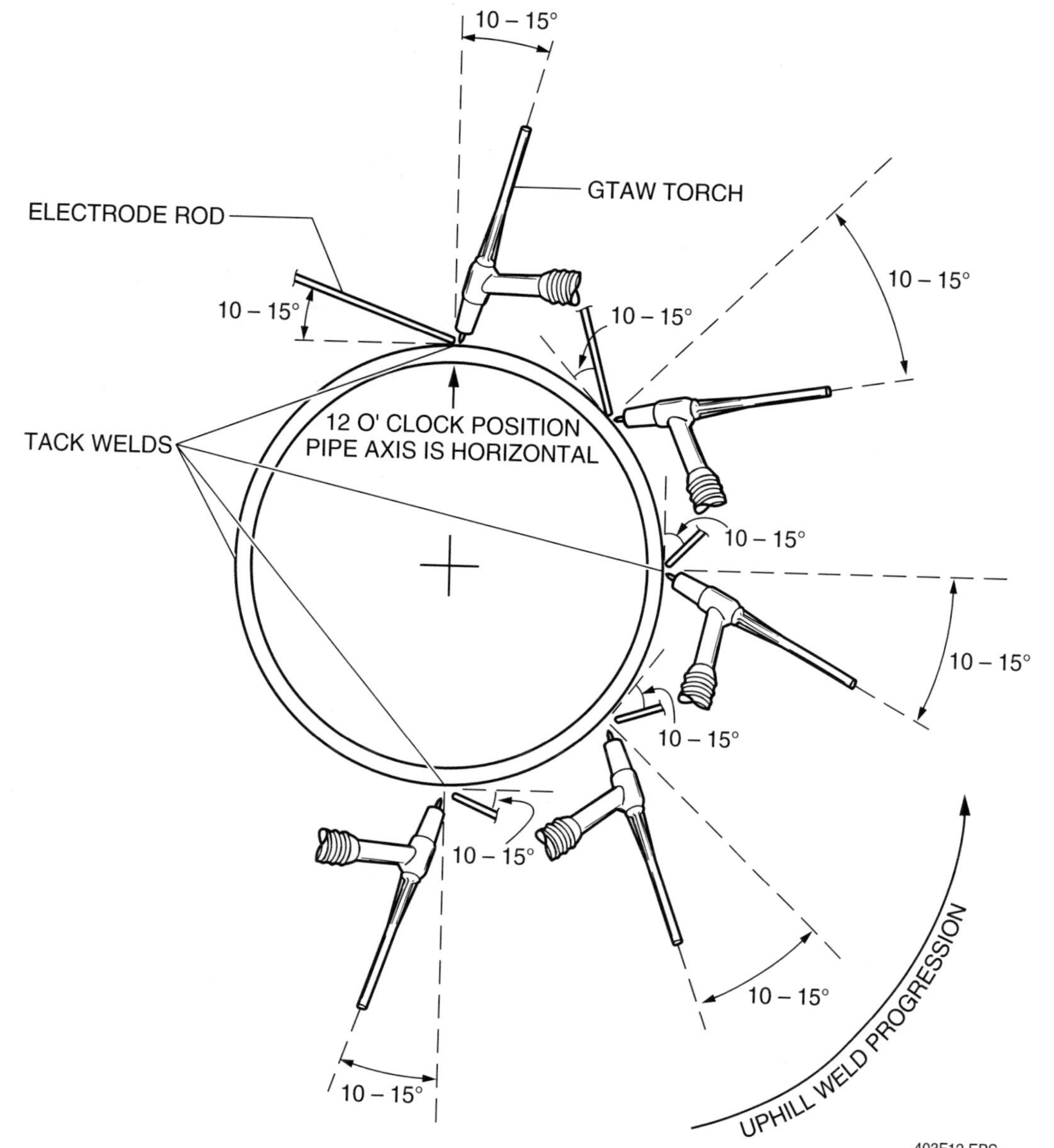

Figure 12 GTAW torch and electrode rod angles for the 5G position.

> **NOTE**
> If required for training purposes, a restricting ring may be added to the 6G position coupon to form a 6GR position coupon.

Follow these steps to practice groove pipe welds in the 6G position:

Step 1 Tack-weld together the practice pipe weld coupon as explained earlier.

Step 2 Clamp or tack-weld the pipe weld coupon into position with the pipe axis inclined 45 degrees to the horizontal plane.

Step 3 Run the root pass uphill with a stringer bead, as shown in *Figure 12*. Repeat the procedure for the opposite side of the pipe.

Step 4 Brush the root pass to clean the weld.

Step 5 Run the remaining passes uphill at the proper work angles. Overlap the passes, and clean the weld after each pass.

MODIFIED U-GROOVE

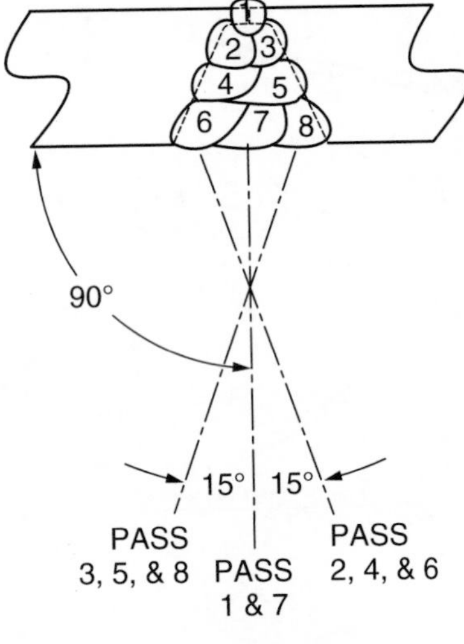

V-GROOVE

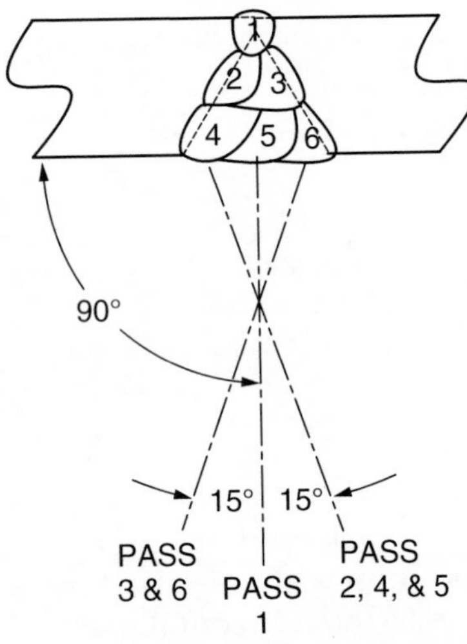

STRINGER BEAD SEQUENCE

403F13.EPS

Figure 13 Multiple-pass 5G bead sequences and work angles.

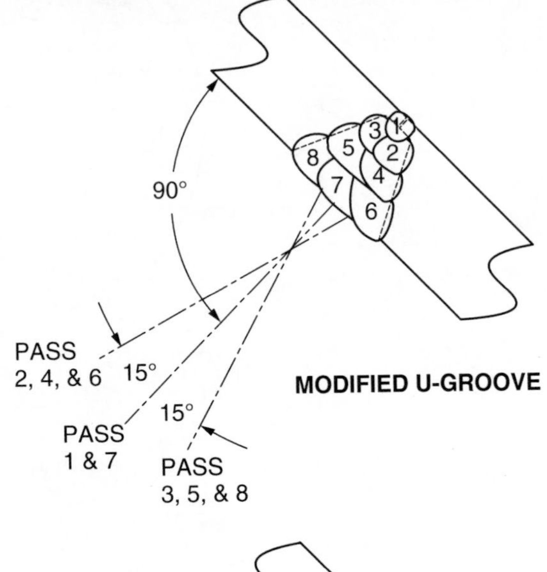

MODIFIED U-GROOVE

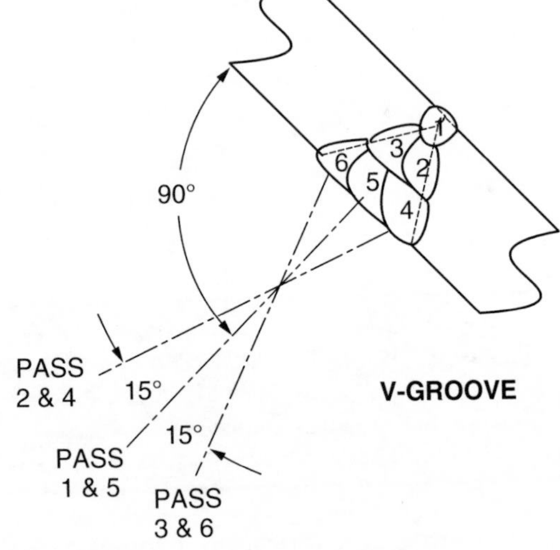

V-GROOVE

STRINGER BEAD SEQUENCE

403F14.EPS

Figure 14 Multiple-pass 6G bead sequences and work angles.

Summary

The ability to make V-groove or modified U-groove welds on aluminum pipe in all positions is one of the more challenging skills you must develop as a welder. The groove weld is the most common weld joint used for joining medium- and thick-walled pipe. You must be able to set up the equipment, perform the welding, and recognize acceptable welds. Groove pipe welds can be made in the 1G-ROTATED, 2G, 5G, and 6G positions. Practice welding in the 2G, 5G, and 6G positions until you can consistently produce acceptable welds.

Review Questions

1. In GTAW of aluminum pipe, the preferred welding table surface is _____.
   a. low-alloy steel
   b. high-carbon steel
   c. aluminum
   d. stainless steel

2. The preferred GTAW joint for aluminum pipe without backing is the _____.
   a. double U-groove weld
   b. double V-groove weld
   c. open-root V-groove weld
   d. modified U-groove weld

3. As the GTAW torch is raised, arc length and voltage _____.
   a. increase, and bead width also increases
   b. decrease, and bead width also decreases
   c. increase, but bead width decreases
   d. decrease, but bead width increases

4. The torch travel angle that is normally used for GTAW of aluminum is a _____.
   a. push angle of 10 to 15 degrees
   b. push angle of 15 to 20 degrees
   c. pull angle of 10 to 15 degrees
   d. pull angle of 15 to 20 degrees

5. The two basic torch and filler metal handling techniques used to perform GTAW are _____.
   a. freehand and on-the-wire
   b. freehand and walking-the-cup
   c. side-to-side and freehand
   d. on-the-wire and side-to-side

6. Modified U-groove and V-groove welds on aluminum pipe are typically used for joining _____.
   a. thin-walled pipe for critical systems
   b. thin-walled pipe for noncritical systems
   c. medium- and thick-walled pipe for critical systems
   d. medium- and thick-walled pipe for noncritical systems

7. GTAW of aluminum pipe usually requires _____.
   a. alternating current with argon shielding gas
   b. alternating current with nitrogen shielding gas
   c. direct current with helium shielding gas
   d. direct current with argon shielding gas

8. The modified U-groove weld is preferred for joints without backing because it _____.
   a. allows for a larger weld pool
   b. prevents complete root penetration
   c. allows for a larger gap in the root opening
   d. reduces the loss of heat from the weld joint

9. To increase penetration during GTAW of aluminum pipe, the shielding gas to use is a mix of _____.
   a. helium and nitrogen
   b. argon and carbon dioxide
   c. helium and carbon dioxide
   d. argon and helium

Review Questions

10. When welding aluminum, the high thermal conductivity of the base metal requires the use of _____.
    a. higher currents
    b. lower currents
    c. higher voltage
    d. lower voltage

11. Pipe groove weld positions are determined by the _____.
    a. axis of the pipe
    b. length of the pipe
    c. weight of the pipe
    d. thickness of the pipe

12. Pipe groove welds without backing should be made with a slight reinforcement *not* to exceed _____.
    a. 1/16"
    b. 1/8"
    c. 1/4"
    d. 1/2"

13. When you are making 2G position groove pipe welds, you should form the weld puddle for the root pass using a(n) _____.
    a. uphill motion
    b. downhill motion
    c. circular motion
    d. side-to-side motion

14. When welding in the 5G position, clamp or tack-weld the pipe weld coupon into position with the pipe axis _____.
    a. vertical
    b. horizontal
    c. inclined at 5 degrees to the horizontal plane
    d. inclined at 45 degrees to the horizontal plane

15. When welding in the 6G position, clamp or tack-weld the pipe weld coupon into position with the pipe axis inclined _____.
    a. 10 degrees to the horizontal plane
    b. 15 degrees to the horizontal plane
    c. 30 degrees to the horizontal plane
    d. 45 degrees to the horizontal plane

Additional Resources

This module is intended to present thorough resources for task training. The following references are suggested for further study. These are optional materials for continued education rather than for task training.

API 1104 – Welding of Pipelines and Related Facilities. Washington, DC: American Petroleum Institute, 2005.

ASME Boiler and Pressure Vessel Code – Section IX: Welding and Brazing Qualifications. New York, NY: ASME International, 2007.

AWS B1.10:1999 Guide for the Nondestructive Examination of Welds. Miami, FL: American Welding Society.

AWS B1.11:2000 Guide for the Visual Examination of Welds. Miami, FL: American Welding Society.

AWS D1.2/D1.2M:2008 Structural Welding Code – Aluminum. Miami, FL: American Welding Society.

AWS D10.11M/D10.11:2007 Guide for Root Pass Welding of Pipe Without Backing. Miami, FL: American Welding Society.

Lincoln Electric website: http://www.lincolnelectric.com offers sources for products and training.

MIG Welding Handbook. Florence, SC: ESAB, 1994.

Modern Welding Technology. Howard B. Cary and Scott Helzer. Englewood Cliffs, NJ: Prentice Hall, Inc., 2005.

OSHA 1910.269, Appendix C, Protection from Step and Touch Potentials. Current edition. Washington, DC: Occupational Safety & Health Administration.

OSHA 1926.351, Arc Welding and Cutting. Current edition. Washington, DC: Occupational Safety & Health Administration.

The Procedure Handbook of Arc Welding. Cleveland, OH: The James F. Lincoln Arc Welding Foundation, 2000.

Welding Aluminum: Theory and Practice. New York, NY: The Aluminum Association, 2002.

Welding Handbook. Volume 1. *Welding Science & Technology.* Miami, FL: American Welding Society, 2001.

Welding Handbook. Volume 2, Part 1: *Welding Processes.* Miami, FL: American Welding Society, 2004.

Welding Pressure Pipelines and Piping Systems. Cleveland, OH: The Lincoln Electric Company, 2000.

Figure Credits

Lincoln Electric Company, Module opener, 403F02, 403F03, 403SA01

Terry Lowe, 403F05 (photo), 403F09, 403F11 (photo), 403F13 (photo), 403F14 (photo), 403A01–403A03 (photos)

Topaz Publications, Inc., 403F06 (photo)

Appendix A

Performance Accreditation Tasks

The Performance Accreditation Tasks (PATs) correspond to and support learning objectives in the *AWS EG2.0:2006 Curriculum Guide for the Training of Welding Personnel; Level I – Entry Welder*.

PATs provide specific acceptable criteria for performance and help to ensure a true competency-based welding program for students.

The following tasks are designed to evaluate your ability to run groove welds with GTAW equipment in three standard test positions using aluminum filler rod of the appropriate diameter and argon shielding gas. Perform each task when you are instructed to do so by your instructor. As you complete each task, show it to your instructor for evaluation. Do not proceed to the next task until told to do so by your instructor. For AWS 2G and 5G certifications, refer to *AWS EG3.0:1996 Guide for the Training and Qualification of Welding Personnel; Level II – Advanced Welder* for bend test requirements. For AWS 6G certifications, refer to *AWS EG4.0:1996 Guide for the Training and Qualification of Welding Personnel; Level III – Expert Welder* for bend test requirements.

V-GROOVE OR MODIFIED U-GROOVE WELDS ON ALUMINUM PIPE IN THE 2G POSITION

As directed by your instructor, use the GTAW process with the appropriate aluminum filler wire, argon shielding gas, and stringer beads to make a multiple-pass groove weld on aluminum pipe in the 2G position with either type of joint.

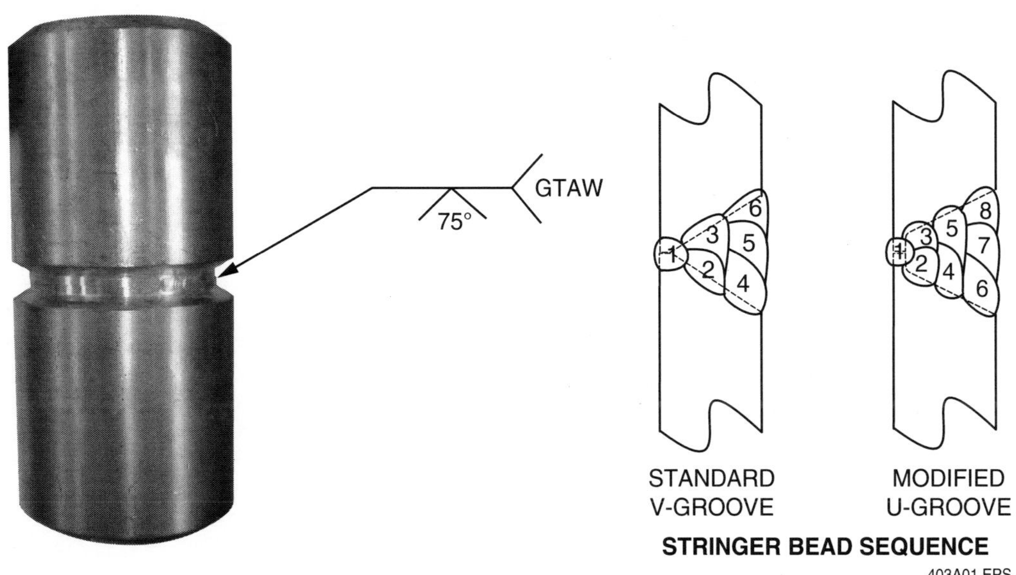

Criteria for Acceptance

- Uniform appearance on the bead face ____________
- Craters and restarts filled to the full cross section of the weld ____________
- Acceptable weld profile in accordance with the *ASME Boiler and Pressure Vessel Code – Section IX* ____________
- Smooth transition with complete fusion at the toes of the weld ____________
- No porosity ____________
- No excessive undercut ____________
- No cracks ____________
- No overlap ____________
- No incomplete fusion ____________

V-GROOVE OR MODIFIED U-GROOVE WELDS ON ALUMINUM PIPE IN THE 5G POSITION

As directed by your instructor, use the GTAW process with the appropriate aluminum filler wire, argon shielding gas, and stringer beads to make a multiple-pass groove weld on aluminum pipe in the 5G position with either type of joint.

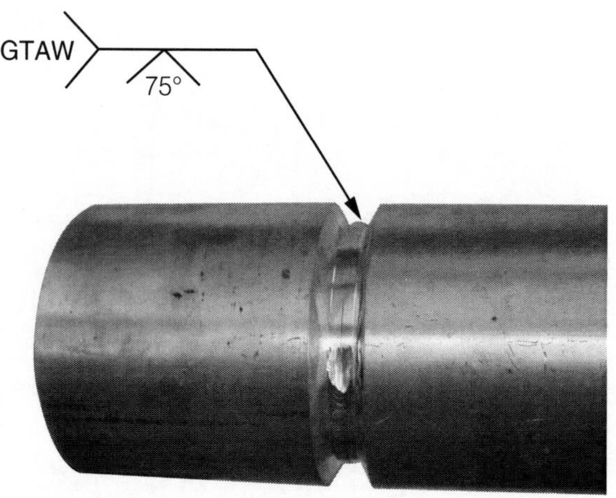

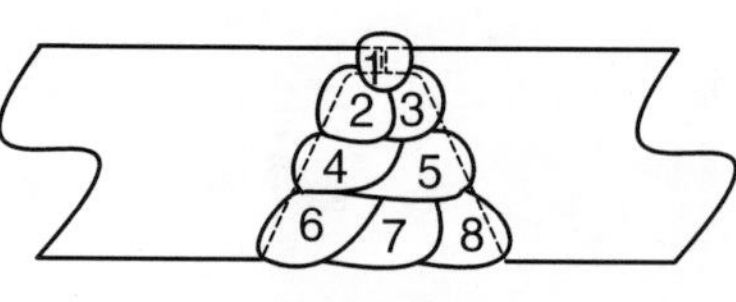

MODIFIED U-GROOVE

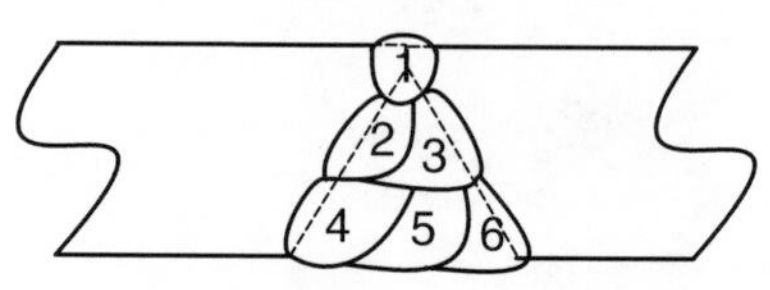

STANDARD V-GROOVE

STRINGER BEAD SEQUENCES

403A02.EPS

Criteria for Acceptance

- Uniform appearance on the bead face __________
- Craters and restarts filled to the full cross section of the weld __________
- Acceptable weld profile in accordance with the *ASME Boiler and Pressure Vessel Code – Section IX* __________
- Smooth transition with complete fusion at the toes of the weld __________
- No porosity __________
- No excessive undercut __________
- No cracks __________
- No overlap __________
- No incomplete fusion __________

V-GROOVE OR MODIFIED U-GROOVE WELDS ON ALUMINUM PIPE IN THE 6G (OR 6GR) POSITION

As directed by your instructor, use the GTAW process with the appropriate aluminum filler wire, argon shielding gas, and stringer beads to make a multiple-pass groove weld on aluminum pipe in the 6G (or 6GR) position with either type of joint.

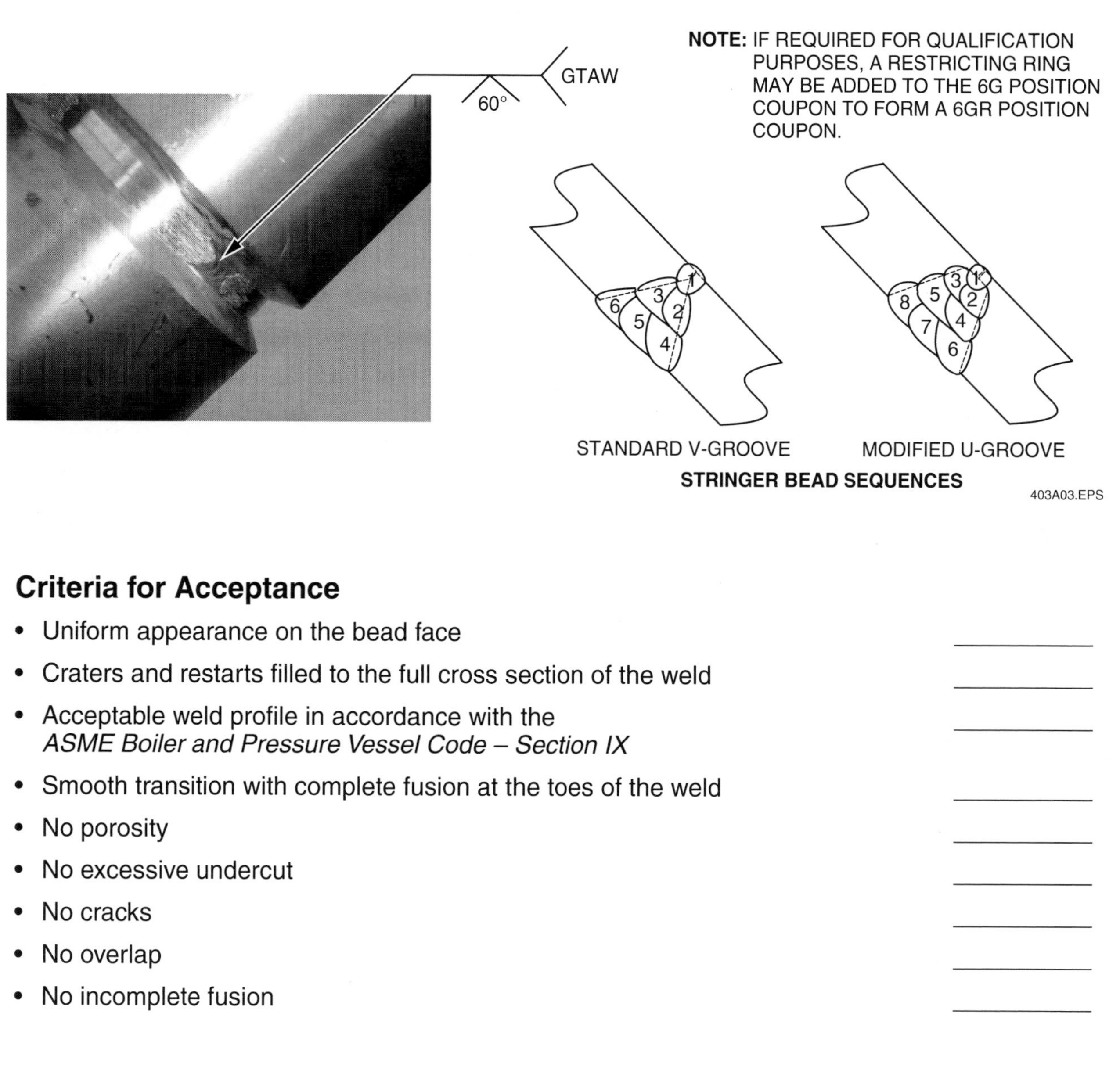

Criteria for Acceptance

- Uniform appearance on the bead face __________
- Craters and restarts filled to the full cross section of the weld __________
- Acceptable weld profile in accordance with the *ASME Boiler and Pressure Vessel Code – Section IX* __________
- Smooth transition with complete fusion at the toes of the weld __________
- No porosity __________
- No excessive undercut __________
- No cracks __________
- No overlap __________
- No incomplete fusion __________

NCCER CURRICULA — USER UPDATE

NCCER makes every effort to keep its textbooks up-to-date and free of technical errors. We appreciate your help in this process. If you find an error, a typographical mistake, or an inaccuracy in NCCER's curricula, please fill out this form (or a photocopy), or complete the online form at **www.nccer.org/olf**. Be sure to include the exact module ID number, page number, a detailed description, and your recommended correction. Your input will be brought to the attention of the Authoring Team. Thank you for your assistance.

Instructors – If you have an idea for improving this textbook, or have found that additional materials were necessary to teach this module effectively, please let us know so that we may present your suggestions to the Authoring Team.

NCCER Product Development and Revision

13614 Progress Blvd., Alachua, FL 32615

Email: curriculum@nccer.org
Online: www.nccer.org/olf

❑ Trainee Guide ❑ AIG ❑ Exam ❑ PowerPoints Other ______________________________

Craft / Level: Copyright Date:

Module ID Number / Title:

Section Number(s):

Description:

Recommended Correction:

Your Name:

Address:

Email: Phone:

GMAW – Aluminum Pipe

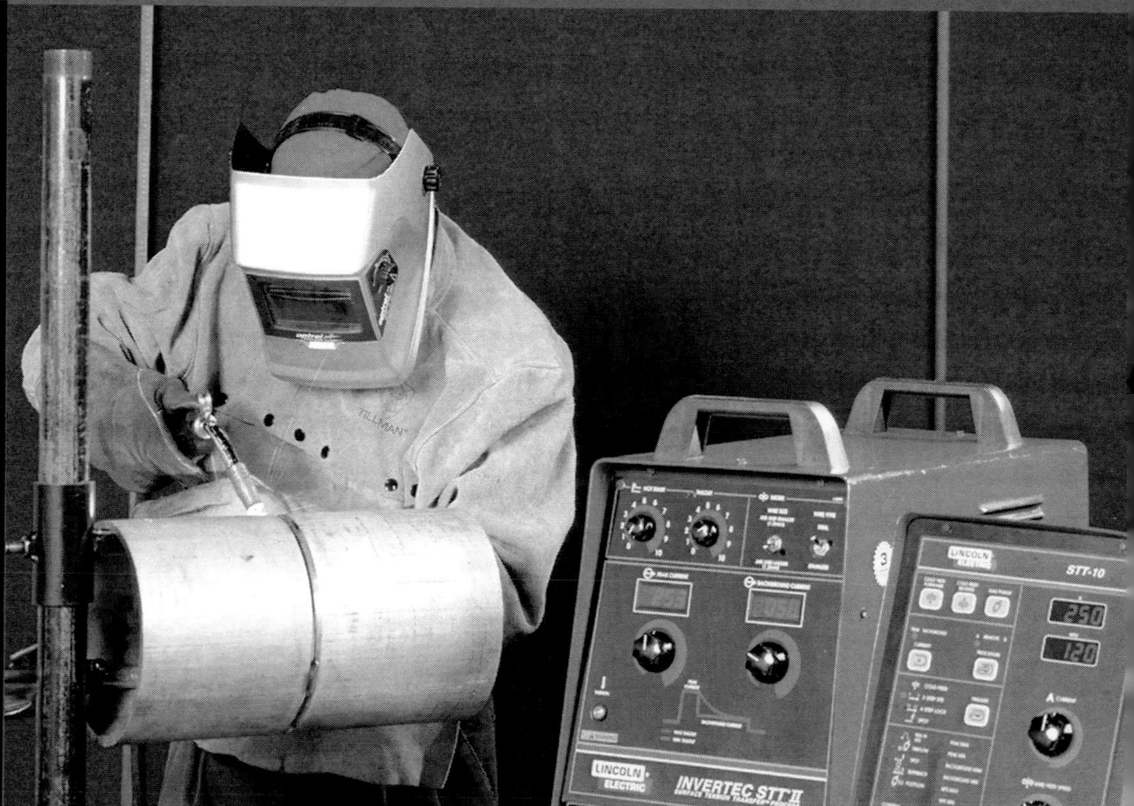

29404-10

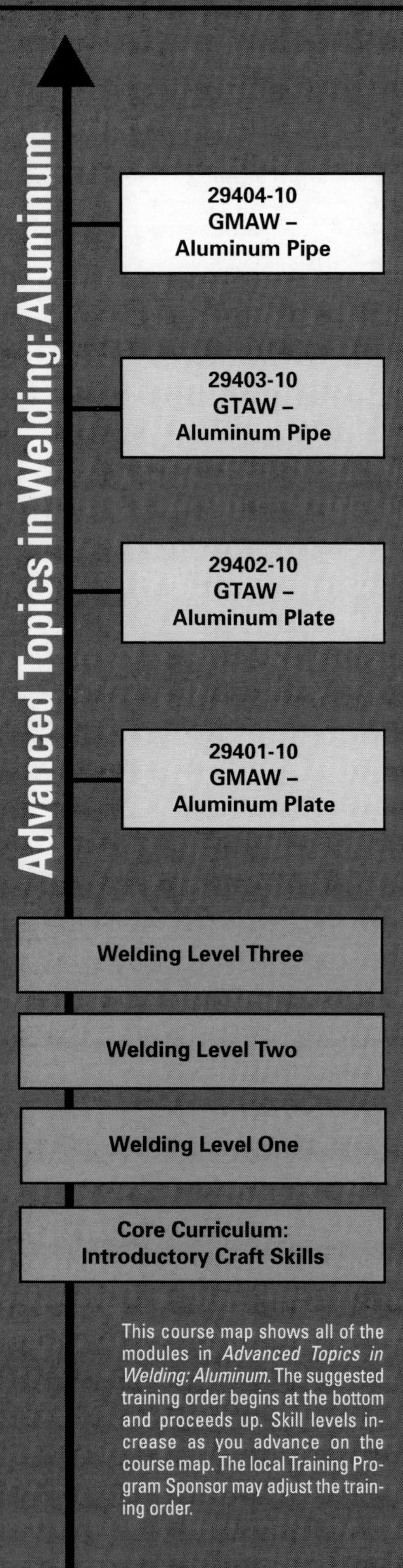

Objectives

When you have completed this module, you will be able to do the following:

1. Explain GMAW preparations associated with making V-groove welds on aluminum pipe.
2. Perform V-groove welds on aluminum pipe with backing in the following positions, using GMAW equipment, aluminum wire, and shielding gas:
   - 2G
   - 5G
   - 6G

Prerequisites

Before you begin this module, it is recommended that you successfully complete *Core Curriculum*; *Welding Level One*; *Welding Level Two*; *Welding Level Three*; and *Advanced Topics in Welding: Aluminum*, Modules 29401-10 through 29403-10.

Contents

Topics to be presented in this module include:

Figures and Tables

1.0.0 INTRODUCTION

GMAW is an arc welding process that uses a continuous, consumable solid wire electrode for the filler metal and shielding gas to protect the weld zone (*Figure 1*). The GMAW process is commonly used to make welds on carbon steel, low-alloy steel, and stainless steel. It is also used for welds on aluminum and other metals. *Figure 2* shows typical GMAW equipment.

GMAW is a fast and effective method for producing high-quality welds. Because this type of welding can be continuous, discontinuities and restarts are reduced. With some materials, such as aluminum, it is a common field practice to use GTAW for the root pass and GMAW to complete the remaining passes. For the purposes of this training module, the dimensions used are representative of various codes and standards and may not be specific to any particular code. Always refer to the applicable code, standard, or site WPS.

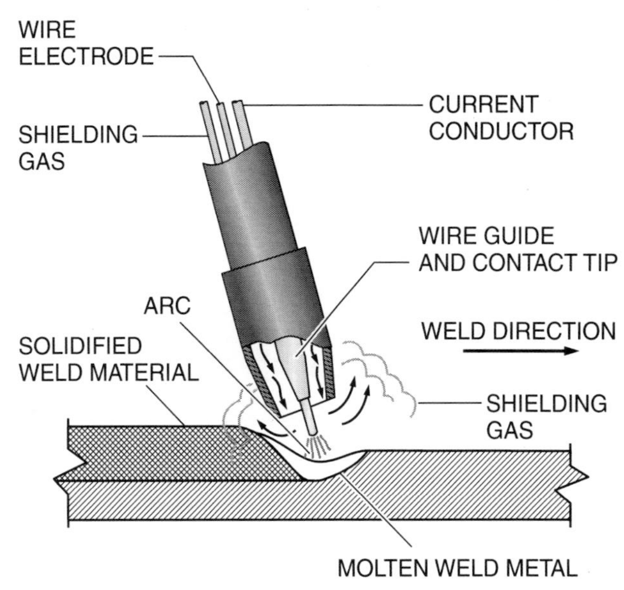

Figure 1 GMAW process.

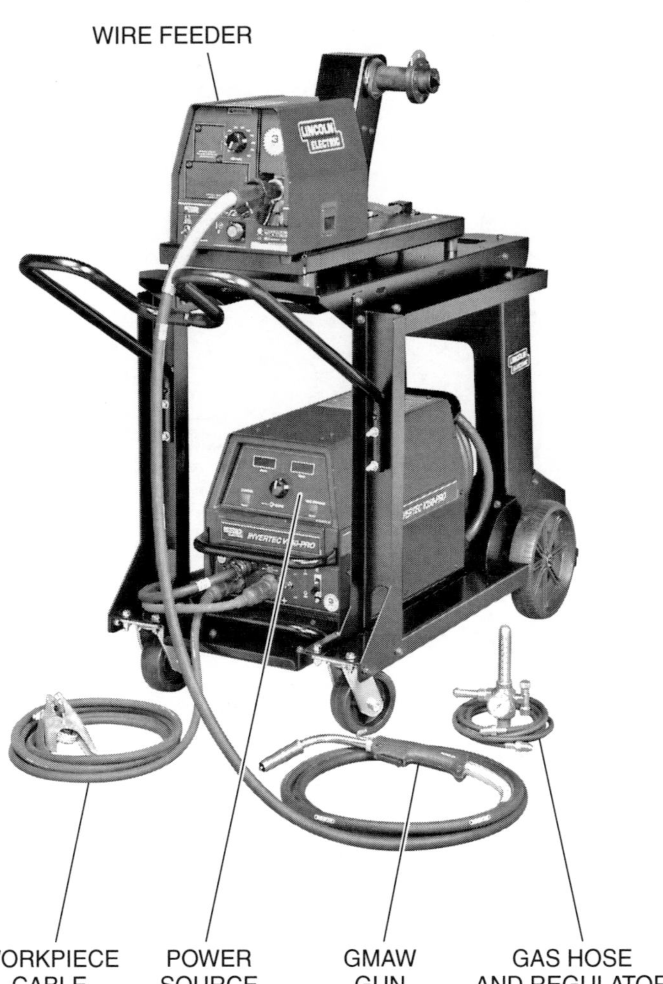

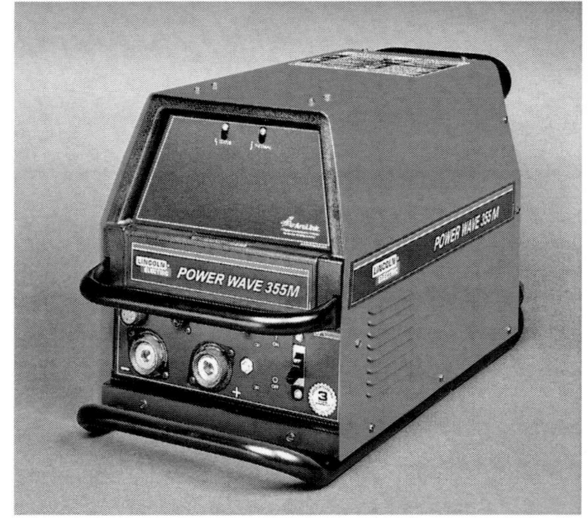

Figure 2 Gas-shielded GMAW equipment.

2.0.0 Safety Summary

The following is a summary of safety procedures and practices that must be observed when welding. Keep in mind that this is only a summary; complete safety coverage is provided in the Level One module *Welding Safety*. If you have not completed that module, do so before continuing. Above all, always use appropriate protective clothing and equipment when welding.

2.1.0 Protective Clothing and Equipment

Welding work creates flying debris, such as sparks or small chunks of hot metal. Anyone welding or assisting a welder must use the proper protective clothing and equipment. The following list provides protective clothing and equipment guidelines:

- Always use safety glasses with a full-face shield or a helmet. The glasses, face shield, or helmet lens must have the proper light-reducing tint for the type of welding being performed. Never directly or indirectly view an electric arc without using a properly tinted lens.
- Wear proper protective leather and/or flame retardant clothing and welding gloves (*Figure 3*). They will protect you from flying sparks, molten metal, and heat.
- Wear 8-inch or taller high-top safety shoes or boots. Make sure that the tongue and lace area is covered by a pant leg. Sometimes the tongue and lace area is exposed or the footwear must be protected from burn marks. In those cases, wear leather spats under the pants or chaps and over the front and top of the footwear.
- Wear a solid material (non-mesh) hat with a bill pointing to the back or toward the ear closest to the welding. This will give added protection. If much overhead welding is required, use a full leather hood with a welding faceplate and the correct tinted lens. If a hard hat is required, use one that allows the attachment of both rear deflector material and a face shield.
- Wear earmuffs or earplugs to protect ear canals from sparks.

2.2.0 Fire/Explosion Prevention

Welding work includes the cutting, grinding, and welding of metal. All of these actions generate heat. They often produce flying sparks. The heat and flying sparks can be the cause of fires and explosions. Welders must use extreme care to pro-

404F03.EPS

Figure 3 Production welder with full protection equipment including respirator.

tect both themselves and others near their work. The following list provides fire and explosion protection guidelines:

- Never carry matches or gas-filled lighters in your pockets. Sparks can cause the matches to ignite or the lighter to explode, causing serious injury.
- Never perform any type of heating, cutting, or welding until a hot work permit has been obtained and an approved fire watch established. Most work-site fires in these types of operations are started by cutting torches.
- Never use oxygen to blow dust or dirt from clothing. The oxygen can remain trapped in the fabric for a time. If a spark hits the clothing during this period, the clothing can burn rapidly and violently out of control.
- Make sure that any flammable material in the work area is either moved to a safe area or shielded by a fire-resistant covering. Approved fire extinguishers must be available before attempting any heating, welding, or cutting operations.
- Always comply with any site requirements for a hot work permit and/or fire watch.
- Never release a large amount of fuel gas, especially acetylene. Methane and propane tend to concentrate in and along low areas. Both gases can ignite at a considerable distance from the release point. Acetylene is lighter than air, but it is even more dangerous than methane. When mixed with air or oxygen, acetylene will explode at much lower concentrations than any other fuel.
- To prevent fires, maintain a neat and clean work area, and make sure that any metal scrap or slag is cold before disposal.

WARNING

Before cutting or welding containers, such as tanks or barrels, check to see if they ever held any explosive, hazardous, or flammable materials. These include, but are not limited to, petroleum products, citrus products, or chemicals that decompose into toxic fumes when heated. As standard practice, always clean and then fill any tanks or barrels with water, or purge them with an appropriate purging gas to displace any oxygen.

2.3.0 Work Area Ventilation

Welders normally work within inches of their welds wearing special protective helmets. Vapors from the welds can be hazardous. The following list provides work area ventilation guidelines:

- Always follow the required confined space procedures before conducting any welding in the confined space.
- Never use oxygen to ventilate confined spaces.

WARNING

An oxygen monitor may be required when working in a confined space.

- Always perform welding operations in a well-ventilated area (*Figure 3*). Welding operations involving zinc or cadmium materials or coatings result in toxic fumes. For long-term welding of these materials, always wear an approved, full-face SAR that uses breathing air supplied from outside of the work area. For occasional, very short-term exposure, you may use a HEPA-rated or metal-fume filter on a standard respirator.
- Make sure confined spaces are properly ventilated for welding operations.

2.4.0 Housekeeping and Fall Protection

Welding work on construction sites is often performed outside and at heights or in close quarters. Welders must keep their work areas as free as possible of tripping hazards. They must also wear all required fall protection equipment when working at heights (*Figure 4*). Make sure to keep all fall protection devices in good condition, untangled, and properly attached. Fall protection equipment may save your life in the event that you react to an electrical shock or are thrown backward after having received one.

2.5.0 Electrical Hazards

As you learned in earlier modules, it takes only a small amount of electrical current to kill you. It takes even less current to make a person flinch or otherwise react to being shocked. That reaction is often what gets a worker seriously injured.

GOING GREEN

Water Disposal

To protect the environment and save resources, make sure to properly dispose of any water used in the cutting of tanks, barrels, or various metals. If the water can be reused, save it and use it again for the next cutting.

404F04.EPS

Figure 4 Working at heights.

Your welding machines or power sources can be powered by either 120 or 480 VAC. As long as the machines are properly wired with all protective shields in place, there should be minimal danger. Turning the ON/OFF switch to the OFF position does not remove the supplied electrical power to the machine. Only experienced electricians may work on electrical arc welding machine power connections. Avoid touching a ground when close to or in contact with a live or hot electrical circuit. The following list provides electrical safety guidelines:

- Check equipment to be sure that it is in good repair.
- Make sure that you are insulated from the workpiece and ground, as well as from other live electrical parts.
- Use plywood, rubber mats, or other dry insulation to stand on. Wear dry gloves that are free of holes.
- Never dip the electrode holder in water to cool it.
- Make sure the equipment is turned off when not in use.
- Electric current flowing through a conductor causes electric and magnetic fields (EMFs). EMFs can interfere with pacemakers and may affect your health in other ways. Consult your physician before arc welding if you have a pacemaker.

To avoid excessive exposure to EMFs, keep the electrode and work cables together, and never place your body between the two cables. Also, never coil the electrode lead around your body, and avoid working directly next to the welding power source.

Another electrical hazard that has been the object of attention is the malfunction of electrical equipment. Electrical shorts, fault protection equipment failures, and accidental back feeds into de-energized equipment can expose workers to potential arc flashes (*Figure 5*). An arc flash event can generate heat up to 35,000°F, melting almost anything within the immediate area.

In *Figure 5*, the electrician is wearing an arc flash suit while working inside an open electrical cabinet. Welders should never be inside electrical cabinets while the electrical power is still applied. However, they may sometimes have to weld on equipment located inside of or very near to high-voltage distribution or process equipment. Under recent changes in the electrical safety guidelines, no unqualified person is allowed within a given distance of a potential arc flash situation. According to the National Fire Protection Association (NFPA), a qualified person is one who has the skills and knowledge related to the construction, operation, and installation of electrical equipment. A qualified person is also one who has received specific safety training on any hazards related to electrical equipment. When in doubt, always check with your local safety representatives before going near electrical distribution equipment.

3.0.0 Welding Preparation

Before welding can begin, the area must be readied, the welding equipment set up, and the metal to be welded prepared. The following sections explain how to set up the equipment for welding.

To practice welding, you will need a welding table, bench, or stand. The welding surface can be steel, but an aluminum surface is preferred. Provisions must be available for placing weld coupons out of position.

Hot Tip

Contact Tube Size for Aluminum Wire

The standard practice is to use a contact tube one size larger than the diameter of the wire.

404F05.EPS

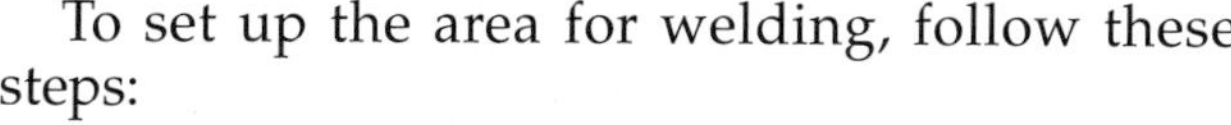

Figure 5 Example of an arc flash from an electrical panel.

To set up the area for welding, follow these steps:

Step 1 Make sure that the area is properly ventilated. Make use of doors, windows, and fans.

Step 2 Check the area for fire hazards. Remove any flammable materials before proceeding.

Step 3 Locate the nearest fire extinguisher. Do not proceed unless the extinguisher is charged and you know how to use it.

Step 4 Set up flash shields around the welding area.

3.1.0 The Welding Machine

Identify a proper welding machine for GMAW, and follow these steps to set it up for use:

Step 1 Verify that the welding machine can be used for GMAW.

Step 2 Know the location of the primary disconnect.

Step 3 Configure the welding machine for GMAW as directed by your instructor (*Figure 6*). Configure the gun polarity (DCEP). Equip the gun with the correct nozzle for the application and with the correct liner material and contact tube for the diameter of the wire being used.

Step 4 In accordance with the manufacturer's instructions, configure and load the wire feeder and gun with solid electrode wire of the proper diameter as directed by your instructor.

CAUTION

To prevent the electrode wire from contaminating the welds, always keep the wire clean when it is in storage or in the machine.

Step 5 Connect the proper shielding gas (argon or argon/helium) for the application as described in the previous level and as specified by the wire electrode manufacturer, WPS, site quality standards, or your instructor.

Step 6 Connect the clamp of the workpiece lead to the workpiece.

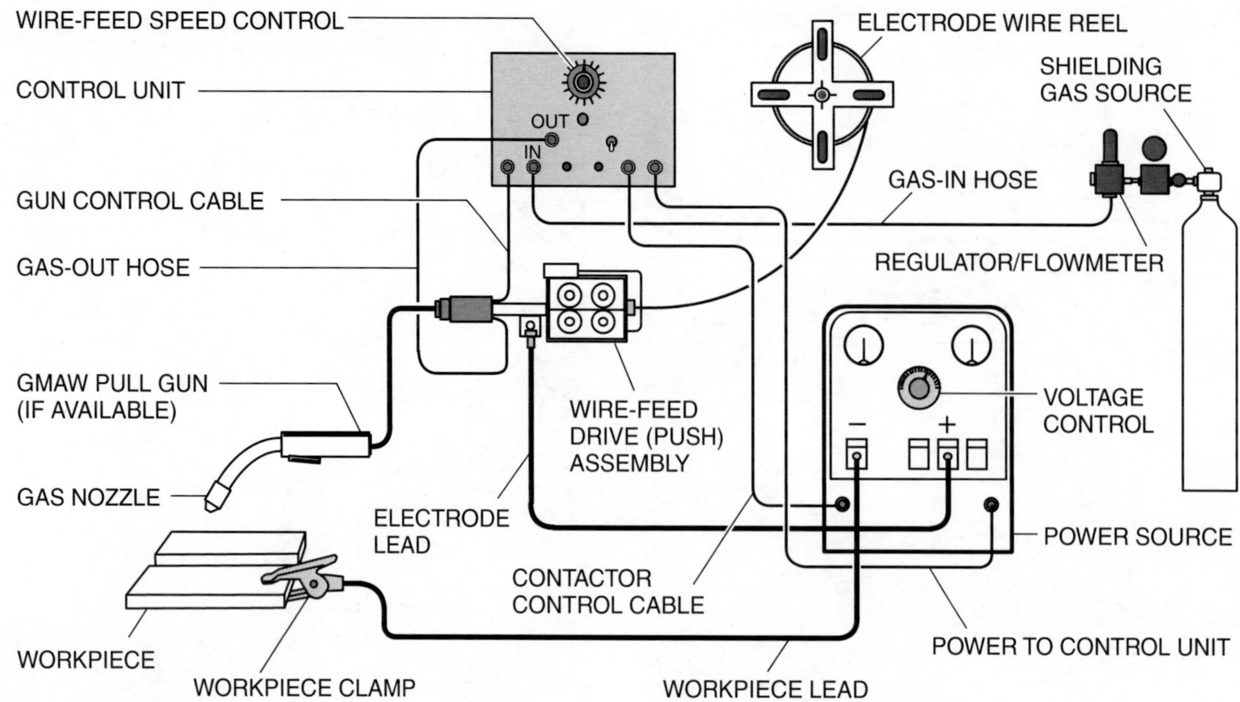

Figure 6 Configuration diagram of a typical GMAW machine.

Step 7 Turn on the welding machine, and purge the gun as directed by the gun manufacturer's instructions.

Step 8 Set the initial welding voltage and wire feed speed as recommended by the manufacturer or the WPS for the type and size electrode wire being used.

3.1.1 Voltage

Arc length is determined by the voltage, which is set at the power source. Arc length is the distance from the wire electrode tip to the base metal or to the molten pool at the base metal (*Figure 7*). If voltage is set too high, the arc will be too long. This can cause the wire to melt and fuse to the contact tube. Voltage that is set too high can also cause porosity and excessive spatter. Voltage must be increased or decreased as the wire feed speed is increased or decreased. Set the voltage to maintain consistent spray transfer.

Hot Tip

Wire Electrode Manufacturer's Recommendations

Always follow the manufacturer's recommendations for using shielding gas and for the initial setting of the welding voltage and wire feed speed parameters. These balanced parameters are critical and are based on the welding position, size, and composition of the solid wire.

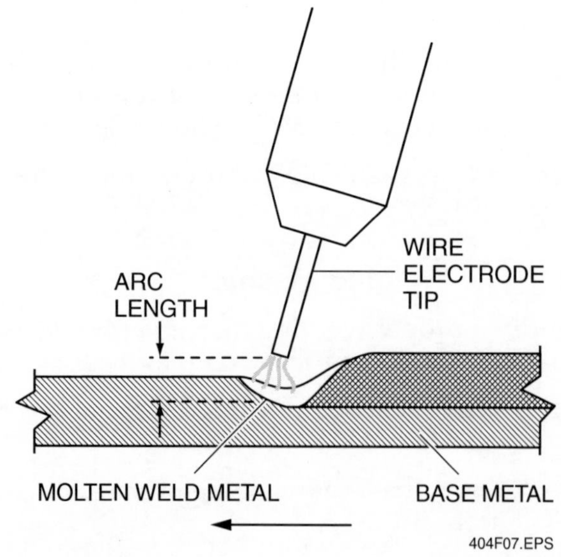

Figure 7 Arc length.

3.1.2 Amperage

With a standard constant-voltage power source, the electrode feed speed controls the welding amperage after the initial recommended setting. The welding power source provides the amperage needed to melt the wire electrode while maintaining the selected welding voltage. Within limits, when the wire electrode feed speed is increased, the welding amperage and deposition rate are also increased. This results in higher welding heat, deeper penetration, and higher beads. When the wire electrode feed speed is decreased, the welding amperage automatically decreases. With lower welding amperage and less heat, the deposition rate drops and the weld beads are smaller with less penetration.

Note that some constant-voltage power sources used for GMAW/FCAW provide varying degrees of current modification. Examples include slope and induction adjustments. Power sources with a slope adjustment allow the welder to vary the amount of amperage change in relation to voltage range of the unit. Standard constant-voltage GMAW/FCAW units have a current slope fixed by the manufacturer for general welding applications and conditions. Pulse transfer power sources allow the peak current for the pulse and the background current between pulses to be adjusted to match specific welding applications and/or wire electrode requirements.

3.1.3 Weld Travel Speed

Weld travel speed is the speed at which the electrode tip passes across the base metal in the forward direction of the weld. It is measured in inches per minute (ipm). Travel speed has a great effect on penetration and bead size. Slower travel speeds build higher beads with deeper penetration. Faster travel speeds build smaller beads with less penetration. Ideally, the welding parameters should be adjusted so that the electrode tip is positioned at the leading edge of the weld puddle during travel.

3.1.4 Gun Position

The gun position in relation to the direction of the weld is defined for aluminum as shown in *Figure 8*. A push angle is used to allow base metal cleaning.

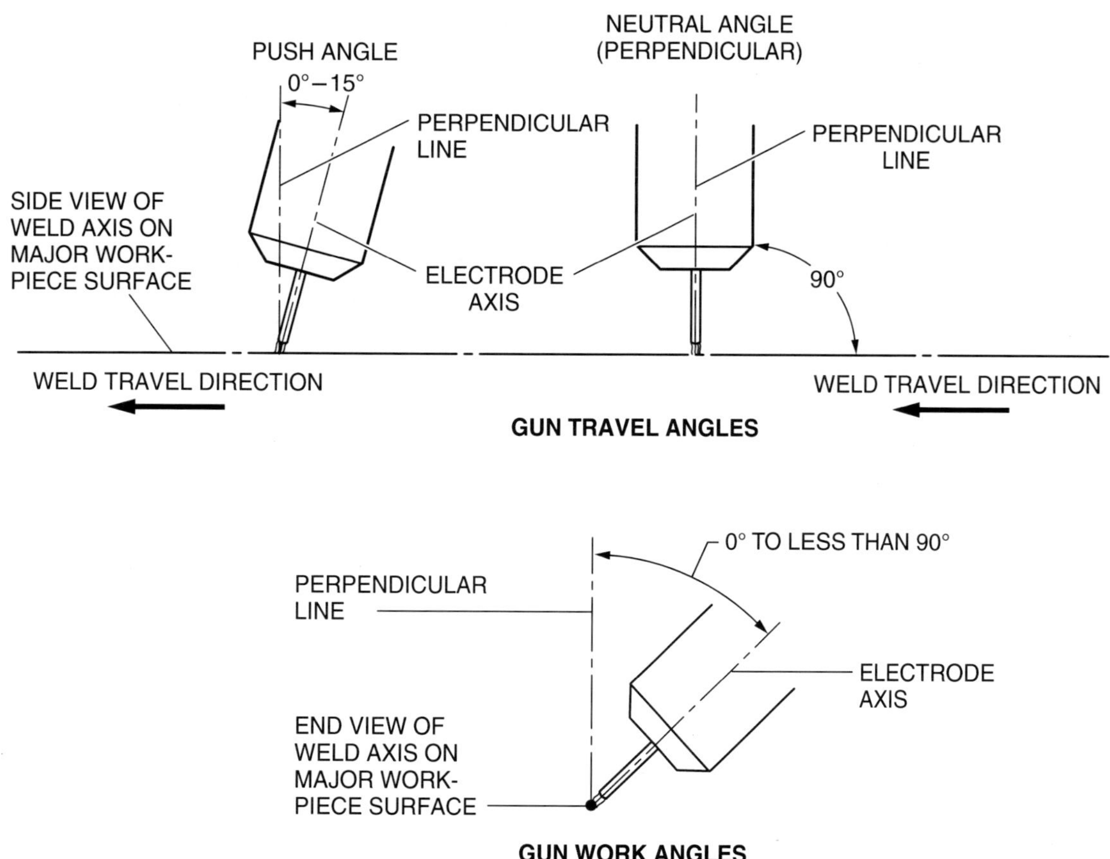

Figure 8 Gun work and travel angles.

Hot Tip

Arc Voltage

A minimum arc voltage is needed to maintain spray transfer. However, penetration is not directly related to voltage. Penetration will increase with voltage for a time, but it will actually decrease if the voltage is increased above its optimum.

3.1.5 Electrode Extension, Stickout, and Standoff Distance

Electrode extension is the length of the wire that extends beyond the tip of the welding gun's contact tube. For high-conductivity metal wires, the preheating effect of wire resistance is minimal. The wire speed and voltage settings have a more direct effect on the amount of weld penetration and deposit rate.

GMAW electrode extension for spray transfer varies from ½" to 1". *Figure 9* shows the typical electrode extension for the spray transfer GMAW gun configuration. It also shows various gun terms and components. Stickout is the distance from the gas nozzle or insulating nozzle to the end of the electrode. Standoff distance is the distance from the gas nozzle or insulating nozzle to the workpiece. Contact tube extension or setback is usually dependent on the transfer mode for the GMAW application. Spray transfer or pulsed spray transfer is normally used for aluminum.

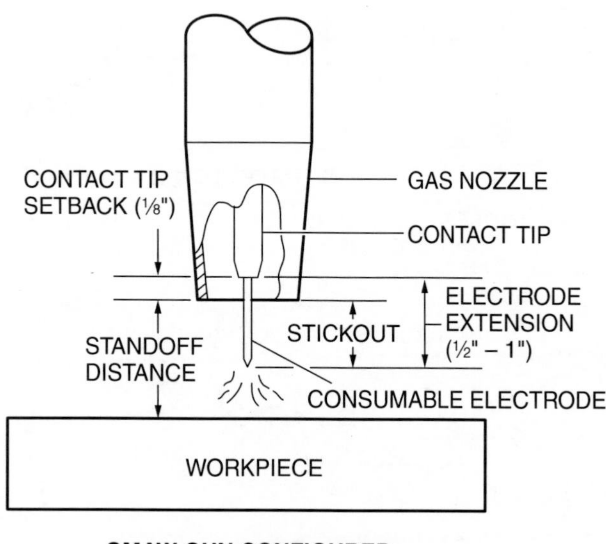

Figure 9 GMAW gun configuration for aluminum welding.

Hot Tip

Travel Speed and Wire Feed Speed

New welders are tempted to turn down the wire feed speed if they have difficulty controlling the weld puddle. Wire electrodes must be run at certain balanced parameters that cannot be changed individually. Voltage, current (if variable), wire feed speed, and travel speed are adjusted and balanced together to control the weld puddle.

3.1.6 Gas Nozzle Cleaning

As the welding machine is used, weld spatter builds up on the gas nozzle and the contact tube. The gas nozzle must be cleaned from time to time with a reamer, round file, or the tang of a standard file. If it is not properly cleaned, it will restrict the shielding gas flow. This will cause porosity in the weld.

3.2.0 Practice Welding Coupons

Pipe weld coupons should be cut from Schedule 40 aluminum pipe that is from 6" to 12" in diameter. Each welded joint requires two coupons of the same size. *Figure 10* shows typical AWS aluminum pipe specifications.

To prepare pipe coupons for V-groove weld joints with backing, follow these steps:

Step 1 Use an approved solvent and a clean rag to remove any grease or oil within a minimum of 1" from the weld joint. Clean the inside and outside of the aluminum pipe with a grinder or stainless steel brush.

GOING GREEN

Conserve Aluminum Practice Coupons

Aluminum for practice welding is expensive and difficult to obtain. To conserve resources and reduce waste in landfills, completely use weld coupons until all surfaces have been welded upon. Cutting coupons apart and reusing the pieces conserves materials. To practice running beads, use any material that cannot be cut into welding coupons.

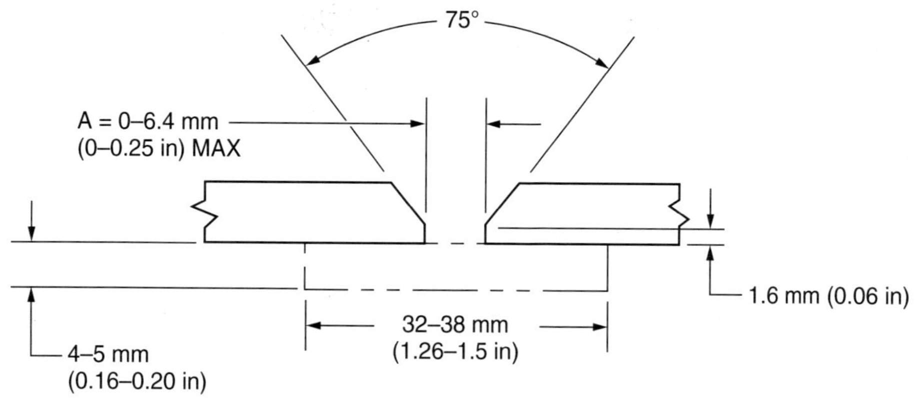

EDGE PREPARATION

PIPING DIMENSIONS							Electrode Diameter		Current DCEP	Argon Flow		Number of Passes[a]
Nominal Pipe Diameter Size Number			Outside Diameter (OD)		Wall Thickness							
DN	NPS	Sch.	mm	(in)	mm	(in)	mm	(in)	amp	L/min	(ft³/h)	A=0
100	4	40	114.3	(4.50)	6.0	(0.24)	1.2	(3/64)	190–210	20	(42)	2
125	5	40	141.3	(5.56)	6.6	(0.26)	1.2	(3/64)	205–225	20	(42)	2
150	6	40	168.3	(6.63)	7.1	(0.28)	1.2	(3/64)	215–235	20	(42)	2
200	8	40	219.1	(8.63)	8.2	(0.32)	1.6	(1/16)	215–235	25	(53)	3
250	10	40	273.1	(10.75)	9.3	(0.37)	1.6	(1/16)	215–235	25	(53)	3
300	12	40	323.9	(12.75)	10.3	(0.41)	1.6	(1/16)	240–260	25	(53)	3

[a] More passes required when A = 6 mm (0.24).
Note: Root opening = 0 for no backing ring or removable backing ring and 6 mm (0.24 in) for permanent backing ring.

404F10.EPS

Figure 10 AWS aluminum pipe groove specifications.

> **NOTE**
>
> If you use a grinding wheel, make sure it is a special high-speed grinding wheel approved for use on aluminum. Make sure that it has not been used on any other metal or used to remove hydrocarbon contaminants.

> **WARNING**
>
> To prevent injury, always comply with the MSDS guidelines for the cleaning solvent(s) being used.

Step 2 Bevel the end of the pipe (*Figure 10*) by any acceptable beveling method. Techniques include mechanical cutting, thermal cutting, or grinding.

Step 3 Cut off a section of the beveled pipe end (4" minimum). For 1G-ROTATED welding, you may need to cut longer coupons to fit the rollers, if used.

Step 4 Check the bevel. There should be no dross. The groove angle should be as shown in *Figure 10*, with no notches deeper than 1/16" (0.06").

Step 5 Grind or file a root face on the bevel as specified by your instructor or as shown in *Figure 10*.

Hot Tip

Machined Grooves and Root Faces

Machining the grooves and root faces of aluminum pipe is the cleanest and most uniform way to prepare pipe for GMAW. This is especially true for aluminum from which all contaminants of thermal cutting must be removed to eliminate weld discontinuities.

> **NOTE**
> For this module, the practice pipe coupons will be tacked with a root opening and a permanent backing ring without the use of any backing gas.

Step 6 Use a permanent backing ring compatible with the size and alloy of the aluminum pipe. Insert the ring inside one of the beveled edges of a pipe coupon section. Make an allowance for the 1⁄4" root opening, and center the backing ring at the joint edge. Clamp the ring to the pipe at several places.

Step 7 Tack-weld the ring to the pipe at four equally spaced positions around the pipe. The tack welds should be narrow and should not exceed 1" in length. Clean the tack welds.

Step 8 Place a beveled edge of the other pipe coupon section over the backing ring. Adjust the section on the ring to allow for the proper root opening. Then make the first of four tacks parallel to one of the existing tack welds on the ring.

Step 9 After the first tack weld, check the root opening on the opposite side. Adjust the root opening if necessary, and make the second tack weld on the opposite side from the first tack.

Step 10 Check the root opening again, and weld the third tack midway between the first two tacks.

Step 11 Weld the fourth tack opposite the third tack and midway between the first and second tacks. There should now be four sets of tack welds evenly spaced (90 degrees) around the pipe coupon (*Figure 11*).

Hot Tip

Welding Large-Diameter Pipe

The practice coupons on large-diameter pipe must be greater than 6" in length. They need more metal to help absorb the increased heat used to make these welds.

Step 12 Clean and feather the tack welds with the edge of a grinding wheel. Feathering the ends of the tack welds with a grinder helps fuse the tack welds into the root pass.

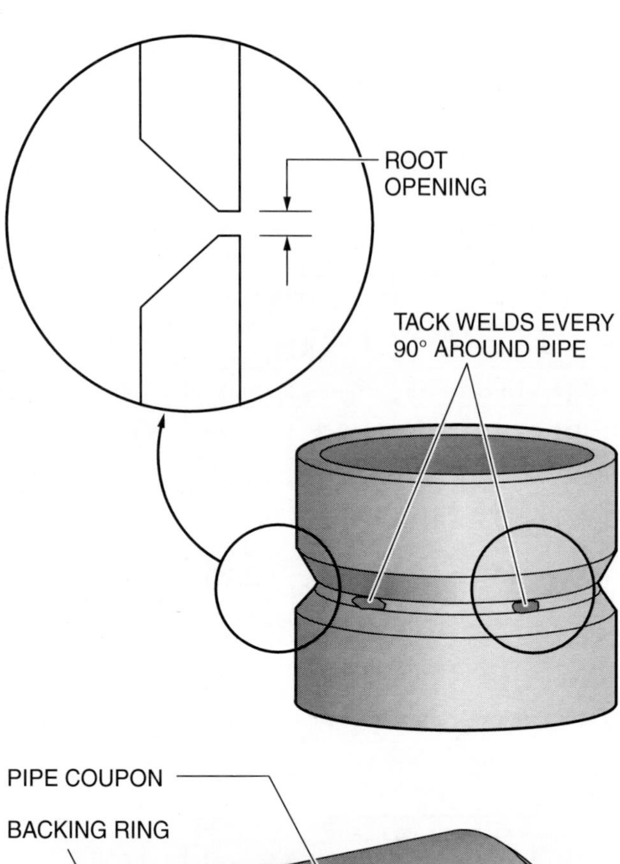

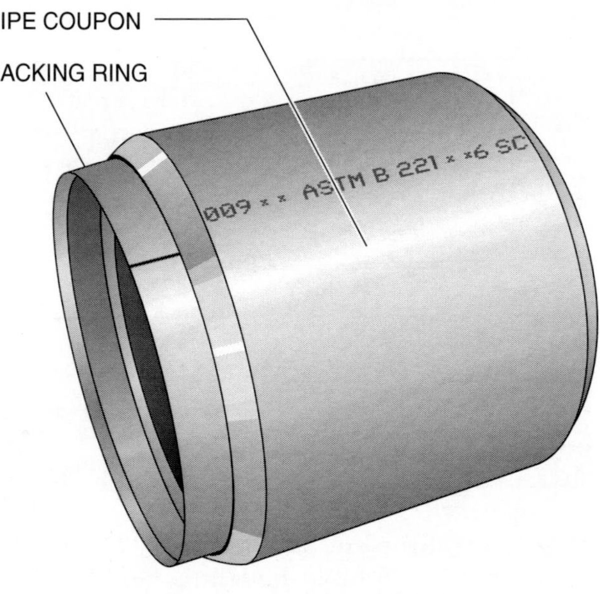

Figure 11 Tacked open-root V-groove weld coupon.

4.0.0 V-Groove Pipe Welds

The V-groove weld with backing is the most common groove weld made on aluminum pipe. These welds are typically used for joining pipe used in critical piping systems. Critical piping systems include pipe, fittings, and welded joints. These systems contain or carry material that has the potential to cause long- or short-term catastrophic danger and damage to people and/or the environment. The damage can occur when components fail to contain or carry the material as designed. Welds in critical piping must meet the most strict code requirements. Noncritical piping is low-pressure piping that is used for heating and air conditioning, simple water systems, and other service installations. The code requirements used to evaluate noncritical piping welds are less strict than those used to assess welds in critical piping systems.

> **CAUTION**
>
> When welding pipe, avoid making arc strikes outside of the weld groove on the surface of the pipe. An arc strike can cause a hardened spot that can crack as the pipe expands and contracts. An arc strike on the pipe's surface is considered a defect, which will require repair or reworking.

Groove welds may be made in all positions on pipe. The weld position is determined by the axis of the pipe. Four standard weld test positions are used with pipe (*Figure 12*).

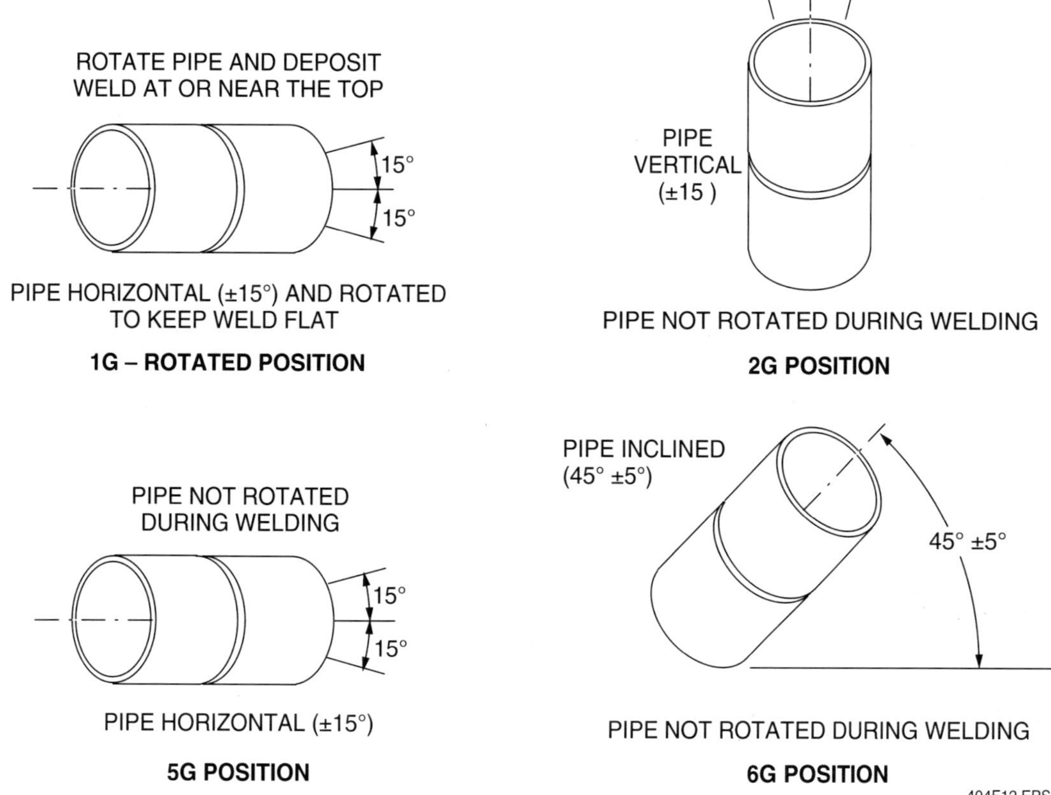

Figure 12 Four basic pipe groove weld test positions.

> **Hot Tip**
>
> **Aluminum Pipe Weld Backing**
>
> Open-root and consumable insert welding do not generally work well with aluminum piping. If you are to make an open-root groove weld, you must back it with a temporary or permanent backing ring; otherwise, you must eliminate the root opening when you use a standard V-groove or modified U-groove joint configuration. No gas backing is required if either of the above methods is used. The modified U-groove joint is recommended for manual AC GMAW.

The sections that follow explain how to perform these V-groove pipe weld positions with backing:

- Flat (1G-ROTATED)
- Horizontal (2G)
- Multiple (5G)
- Multiple inclined (6G)

4.1.0 Practicing Flat (1G-ROTATED) Position V-Groove Welds

Practice the 1G-ROTATED position V-groove pipe welds with backing. Use a shielding gas and aluminum filler wire that has a diameter specified by your instructor.

For the root pass, keep the gun angle at 90 degrees (0-degree work angle) to the pipe axis with a 10- to 15-degree push angle. Use a slight side-to-side oscillation to control penetration. Be sure to fuse into the tack welds. Clean the face of the root pass with a brush.

When running the remaining passes, use stringer or weave beads with a slight side-to-side motion to ensure fusion at the toes of the weld bead. Try to keep the wire at the leading edge of the weld puddle to ensure proper penetration. Take special care to fill the crater at the termination of the weld. Run all passes at or near the top of the pipe as it is rotated.

Follow these steps to practice V-groove pipe welds with backing in the 1G-ROTATED position:

Step 1 Tack-weld together the practice pipe weld coupon as explained earlier.

Hot Tip

Test Position

Welding pipe in the 6G position is a common welding test. A ring may be added to test for restricted accessibility (6GR).

Fill Pass

When running the fill pass, fill the craters to prevent cracking, and stagger the location of the starts and stops for each weld.

Step 2 Position the pipe weld coupon horizontally on two sets of rollers at a comfortable welding height. *Figure 13* shows roller supports commonly found in pipe welding shops.

Step 3 Make sure that the workpiece clamp is attached directly to the pipe coupon. This will prevent the welding current from passing through the roller bearings or arcing between the rollers and the pipe coupon.

CAUTION

Failure to attach the workpiece clamp to the pipe coupon can result in variations in welding current, damage to the roller bearings, and arcing on the rollers and pipe coupon.

Step 4 Run the root pass using a slight side-to-side oscillation. Place a tack weld at the 11 o'clock position. Start the weld bead on the tack weld, and advance toward the 12 o'clock position.

Step 5 Roll the pipe as necessary to keep the weld in the flat position.

Step 6 Clean the weld.

Step 7 Use the same rolling procedure to make the remaining passes. Use stringer or weave beads as applicable. Pay particular attention to ensure proper fusion and prevent excess buildup. Overlap the passes (*Figure 14*), and clean the weld after each pass.

4.2.0 Practicing Horizontal (2G) Position V-Groove Welds

Practice making V-groove pipe welds with backing in the 2G position. Use a shielding gas and an aluminum filler wire that has a diameter specified by your instructor.

For the root pass, the gun should be at a 10- to 15-degree push angle and at 90 degrees (0-degree work angle) to the surface of the pipe. To prevent the weld puddle from sagging, the gun work angle can be dropped slightly, but not more than 10 degrees. Use a slight side-to-side oscillation to control penetration (*Figure 15*). Be sure to fuse into the tack welds.

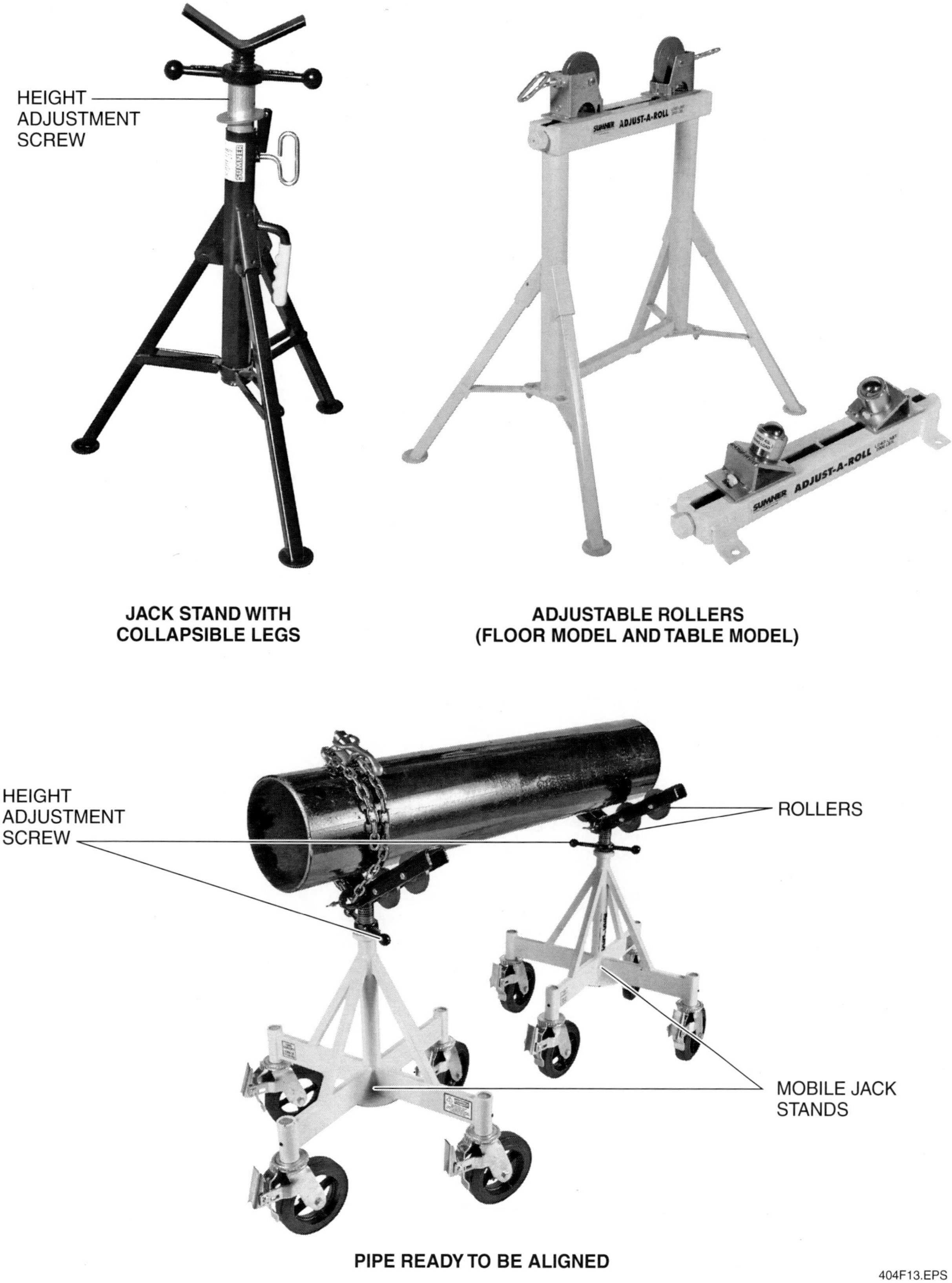
HEIGHT ADJUSTMENT SCREW
JACK STAND WITH COLLAPSIBLE LEGS
ADJUSTABLE ROLLERS (FLOOR MODEL AND TABLE MODEL)
HEIGHT ADJUSTMENT SCREW
ROLLERS
MOBILE JACK STANDS
PIPE READY TO BE ALIGNED
404F13.EPS

Figure 13 Pipe roller supports.

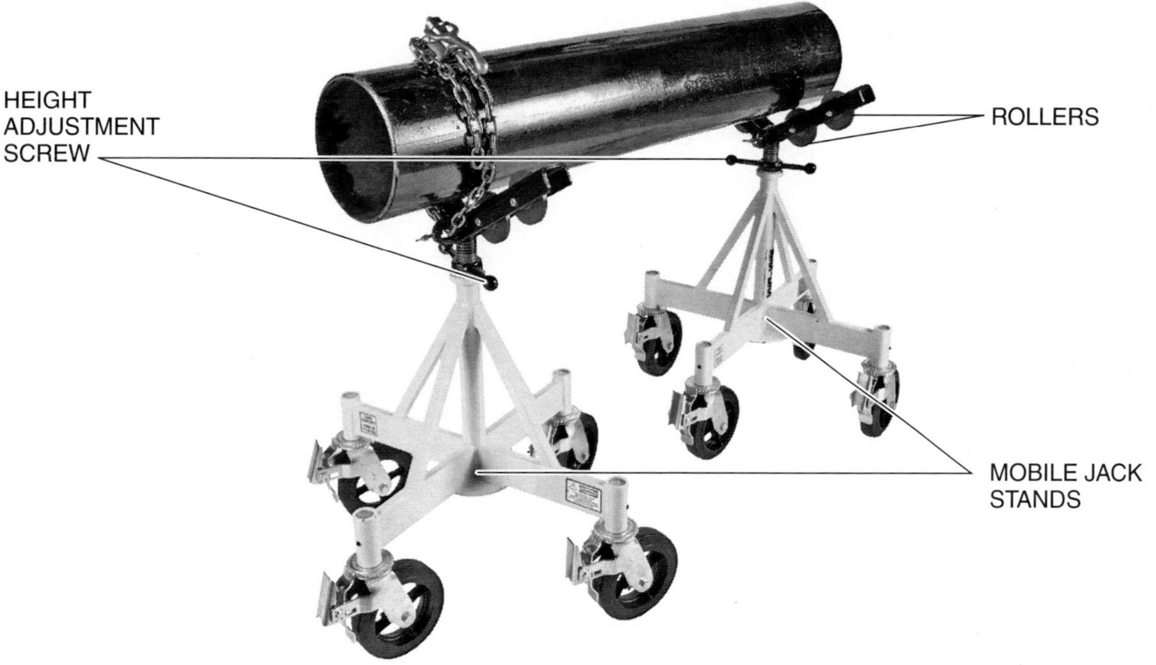

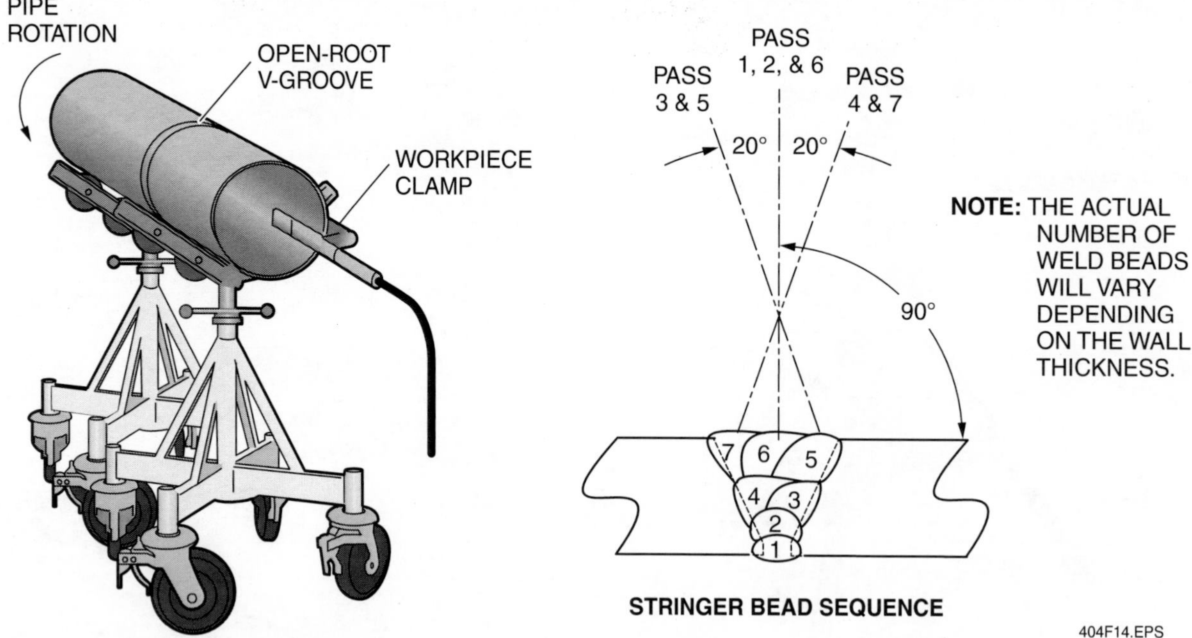

Figure 14 Multiple-pass 1G-ROTATED bead sequence and work angles.

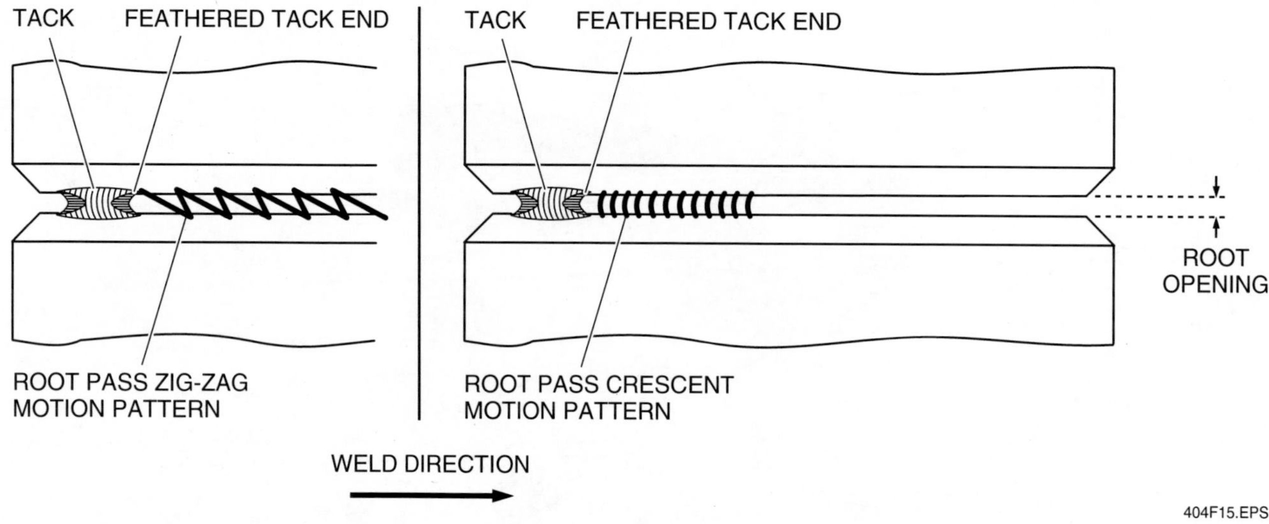

Figure 15 Root pass motion pattern.

Hot Tip

Powered Pipe Roller

A pipe roller that is operated by an electric motor activated by a foot switch is shown here. These devices are primarily used for large, heavy pipe sections.

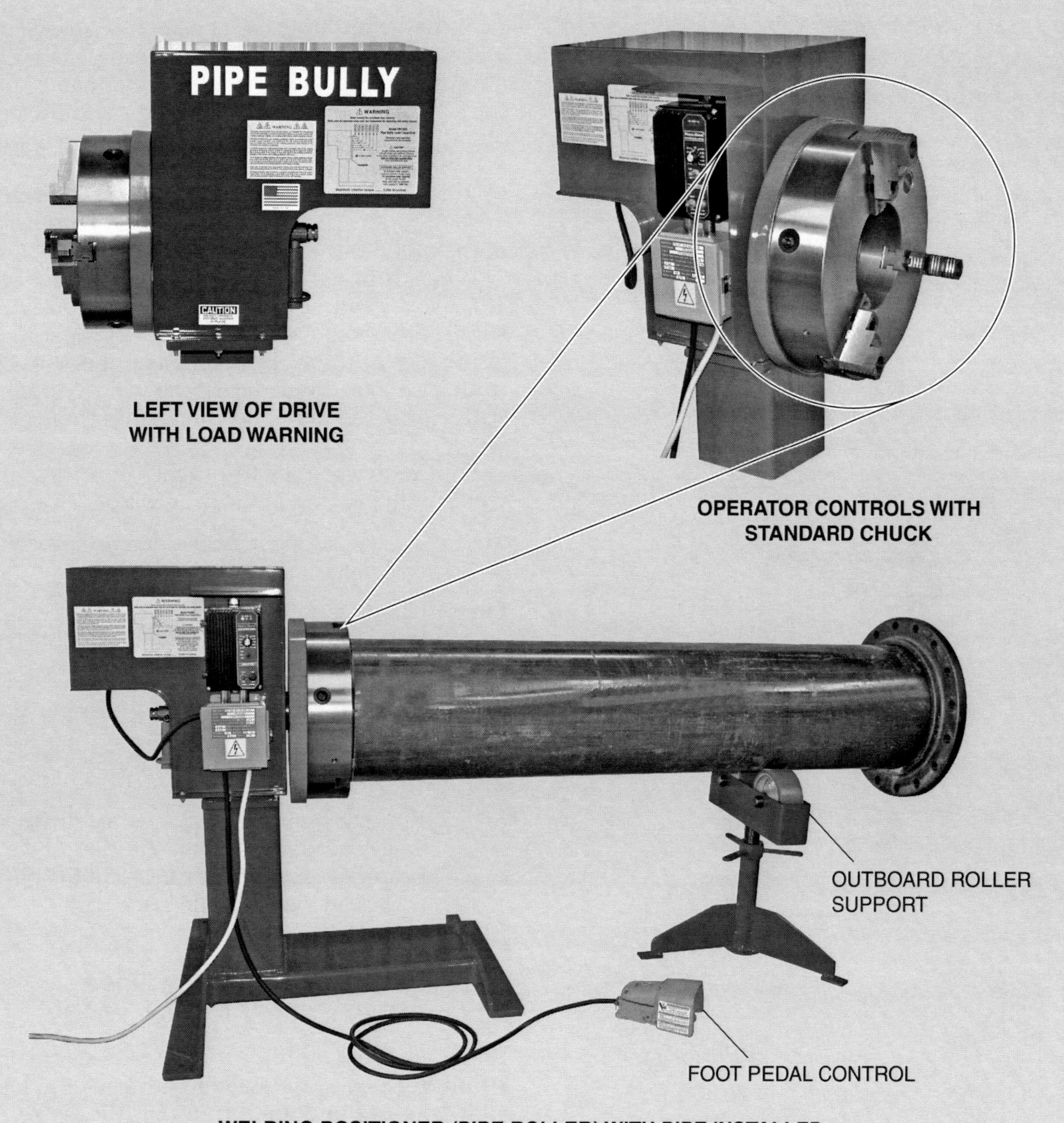

WELDING POSITIONER (PIPE ROLLER) WITH PIPE INSTALLED

404SA01.EPS

When running the remaining passes (*Figure 16*), use stringer beads and a slight side-to-side motion to ensure proper fusion at the toes of the weld bead. Try to keep the wire electrode at the leading edge of the weld puddle. Take special care to fill the crater at the termination of the weld

Follow these steps to practice V-groove pipe welds with backing in the 2G position:

Step 1 Tack-weld together the practice pipe weld coupon as explained earlier.

Step 2 Clamp or tack-weld the pipe coupon with the axis of the pipe vertical.

Step 3 Starting on a tack weld, run the root pass as shown in *Figure 16.* Clean the crater between any necessary restarts. Continue this procedure until the root pass has been completed.

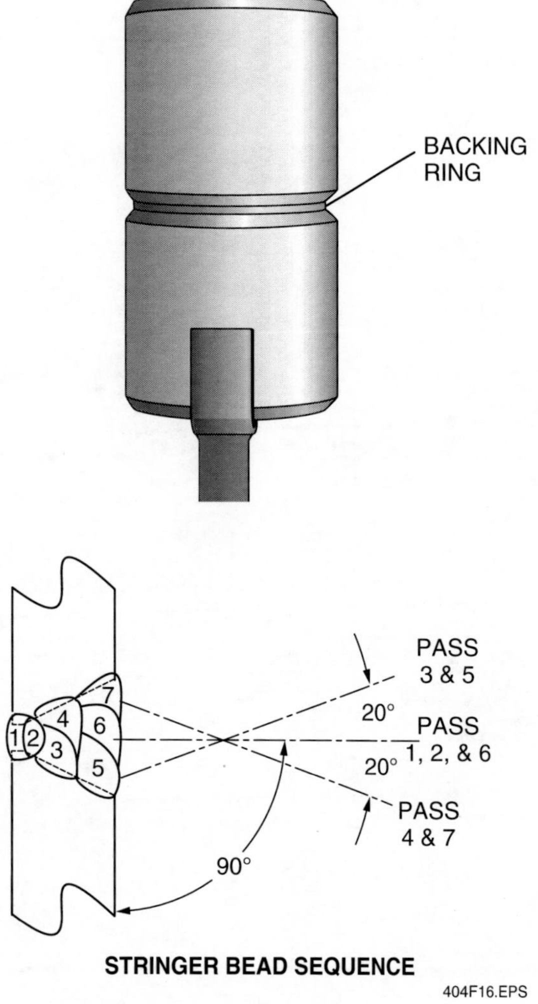

Figure 16 Multiple-pass 2G bead sequence and work angles.

Step 4 Clean the weld.

Step 5 Run the remaining passes at the appropriate work angles. Overlap the passes, and clean the weld after each pass.

4.3.0 Practicing Multiple (5G) Position V-Groove Welds

Practice the multiple (5G) position V-groove pipe welds with backing. Use a shielding gas and an aluminum filler wire that has a diameter specified by your instructor. The gun should be at about a 10- to 15-degree push angle for the root and all remaining passes. All welds will be run as uphill passes. *Figure 17* shows the travel positions needed for the passes that are made on both sides of the pipe.

When running the remaining passes (*Figure 18*), use weave beads or stringer beads with a slight side-to-side motion to ensure proper fusion at the toes of the weld bead. Keep the wire electrode at the leading edge of the weld puddle. Be sure to fill the crater at the termination of the weld.

Follow these steps to practice V-groove pipe welds with backing in the 5G position:

Step 1 Tack-weld together the practice pipe weld coupon as explained earlier.

Step 2 Clamp or tack-weld the pipe weld coupon into position with the pipe axis horizontal.

Step 3 Run the root pass with a stringer bead, modifying the travel angles as necessary.

Step 4 Clean the weld.

Step 5 Run the remaining passes as stringer or weave beads in an uphill direction at the appropriate work angles. Overlap the passes, and clean the weld after each pass.

4.4.0 Practicing Multiple Inclined (6G) Position V-Groove Welds

Practice the 6G multiple inclined (45-degree) position V-groove pipe welds with backing. Use a shielding gas and an aluminum filler wire that has a diameter specified by your instructor. The gun should be at about a 10- to 15-degree push angle for the root and all remaining passes. All welds will be run as uphill passes. *Figure 17* shows the travel positions needed for the passes that are made on both sides of the pipe.

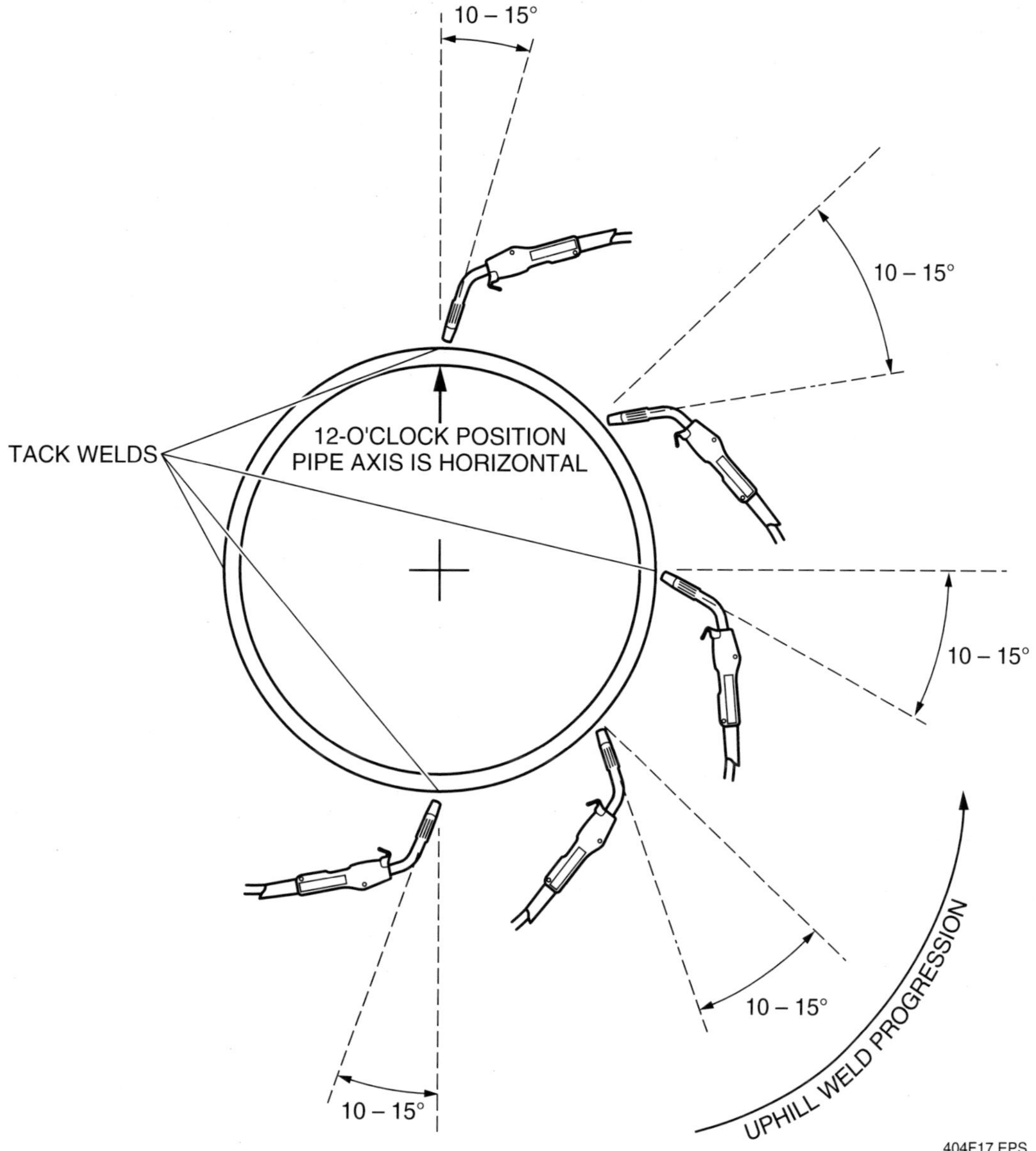

Figure 17 Multiple-pass weld progression for the 5G position.

When running the remaining passes, use weave beads or stringer beads with a slight side-to-side motion to ensure fusion at the toes of the weld bead. Keep the wire electrode at the leading edge of the weld puddle. Be sure to fill the crater at the termination of the weld. *Figure 19* shows the stringer bead sequence and work angles for the 6G position.

Follow these steps to practice V-groove pipe welds with backing in the 6G position:

> **NOTE**
> If required for training purposes, a restricting ring may be added to the 6G position coupon to form a 6GR position coupon.

Step 1 Tack-weld together the practice pipe weld coupon as explained earlier.

Step 2 Clamp or tack-weld the pipe weld coupon into position with the pipe axis inclined 45 degrees to the horizontal plane.

Step 3 Run the root pass with a stringer bead, modifying the travel angles as necessary.

Step 4 Clean the weld.

Step 5 Run the remaining passes in an uphill direction at the appropriate work angles. Overlap the passes, and clean the weld after each pass.

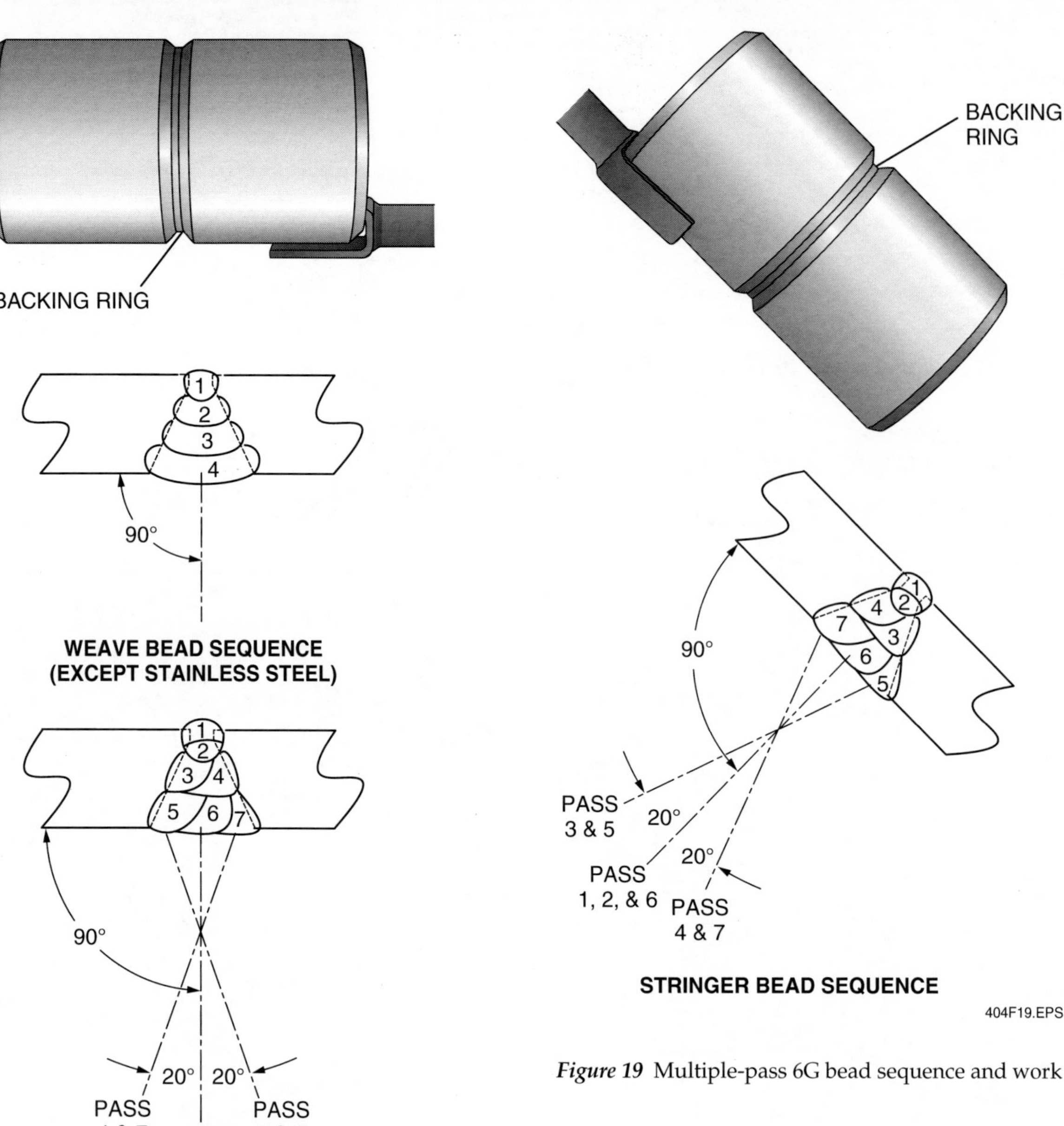

Figure 18 Multiple-pass 5G bead sequences and work angles.

Figure 19 Multiple-pass 6G bead sequence and work angles.

Summary

This module covered GMAW safety, including fall protection and electrical safety. It also provided information about setting up GMAW equipment, preparing the welding work area, and making acceptable V-groove welds with backing on aluminum pipe in all positions.

The V-groove weld is the most common weld joint used for joining pipe used in critical piping systems. For aluminum pipe welding, you must be able to set up the equipment, prepare the coupons for welding, recognize acceptable welds, and make V-groove pipe welds in all applicable positions. For the purposes of this module, practice making V-groove welds on aluminum pipe in the 2G, 5G, and 6G positions until you can consistently produce acceptable welds.

Review Questions

1. For the filler metal, the GMAW process for aluminum uses _____.
   a. a continuous solid wire electrode
   b. nickel-coated aluminum rods
   c. short lengths of electrode
   d. tin-coated aluminum rods

2. When welding aluminum, GTAW is often used for the root pass, and the process commonly used for the remaining passes is _____.
   a. SMAW
   b. FCAW-G
   c. GMAW
   d. FCAW-S

3. Electric current flowing through a conductor causes electric and magnetic fields (EMFs) that can interfere with _____.
   a. vision
   b. breathing
   c. pacemaker operation
   d. normal heart function

4. When welding amperage is increased, the result is _____.
   a. deeper penetration
   b. less penetration
   c. lower beads
   d. lower heat

5. The GMAW application normally used for aluminum is spray transfer or _____.
   a. globular transfer
   b. deposit transfer
   c. short-circuiting transfer
   d. pulsed spray transfer

6. Contact tube extension or setback is usually dependent on the _____.
   a. transfer mode for the GMAW application
   b. distance from the gun nozzle to the workpiece
   c. amperage
   d. voltage

7. If the weld spatter that accumulates in the gas nozzle is not removed, it will restrict the shielding gas flow to the weld and cause _____.
   a. cold fusion
   b. porosity
   c. cracking
   d. worm holes

8. To clean weld spatter from a gas nozzle, you can use a _____.
   a. stainless steel brush
   b. chemical cleaner
   c. round file
   d. grinder

9. Pipe weld coupons should be cut from Schedule 40 aluminum pipe that has a diameter of _____.
   a. 1" to 3"
   b. 6" to 12"
   c. 3" to 18"
   d. 12" to 15"

10. When preparing pipe coupons for V-groove weld joints with backing, tack-weld a permanent backing ring to the pipe at _____.
    a. two equally spaced positions
    b. four equally spaced positions
    c. six equally spaced positions
    d. eight equally spaced positions

11. The most common groove weld made on aluminum pipe is the _____.
    a. single-groove weld
    b. bevel-groove weld
    c. open-root V-groove weld
    d. V-groove weld with backing

12. When welding pipe, avoid making an arc strike outside the weld groove on the surface of the pipe because it can cause a _____.
    a. hardened spot that can crack
    b. keyhole that must be filled
    c. thick, rough area that must be filed
    d. thin, weak spot that might expand

Review Questions

13. To ensure proper penetration when welding aluminum pipe in the 1G-ROTATED position, keep the filler metal _____.
    a. just above the weld puddle
    b. at the trailing edge of the weld puddle
    c. at the leading edge of the weld puddle
    d. at the center of the weld puddle

14. For running the remaining passes when welding aluminum pipe in the 5G position, hold the gun using a push angle of _____.
    a. 15 to 20 degrees
    b. 10 to 15 degrees
    c. 0 to 5 degrees
    d. 0 degrees

15. When running the remaining passes in the 6G position, you can ensure fusion at the toes of the weld bead by using a _____.
    a. front-to-back motion
    b. side-to-side motion
    c. downhill motion
    d. circular motion

Additional Resources

This module is intended to present thorough resources for task training. The following references are suggested for further study. These are optional materials for continued education rather than for task training.

API 1104 – Welding of Pipelines and Related Facilities. Washington, DC: American Petroleum Institute, 2005.

ASME Boiler and Pressure Vessel Code – Section IX: Welding and Brazing Qualifications. New York, NY: ASME International, 2007.

AWS B1.10:1999 Guide for the Nondestructive Examination of Welds. Miami, FL: American Welding Society.

AWS B1.11:2000 Guide for the Visual Examination of Welds. Miami, FL: American Welding Society.

AWS C5.6-1989 Recommended Practices for Gas Metal Arc Welding. Miami, FL: American Welding Society.

AWS D1.2/D1.2M:2008 Structural Welding Code – Aluminum. Miami, FL: American Welding Society.

AWS D10.7M/D10.7:2008 Guide for the Gas Shielded Arc Welding of Aluminum and Aluminum Alloy Pipe. Miami, FL: American Welding Society.

Lincoln Electric website: http://www.lincolnelectric.com offers sources for products and training.

MIG Welding Handbook. Florence, SC: ESAB, 1994.

Modern Welding Technology. Howard B. Cary and Scott Helzer. Englewood Cliffs, NJ: Prentice Hall, Inc., 2005.

OSHA 1910.269, Appendix C, Protection from Step and Touch Potentials. Current edition. Washington, DC: Occupational Safety & Health Administration.

OSHA 1926.351, Arc Welding and Cutting. Current edition. Washington, DC: Occupational Safety & Health Administration.

The Procedure Handbook of Arc Welding. Cleveland, OH: The James F. Lincoln Arc Welding Foundation, 2000.

Welding Aluminum: Theory and Practice. New York, NY: The Aluminum Association, 2002.

Welding Handbook. Volume 1. *Welding Science & Technology.* Miami, FL: American Welding Society, 2001.

Welding Handbook. Volume 2, Part 1: *Welding Processes.* Miami, FL: American Welding Society, 2004.

Welding Pressure Pipelines and Piping Systems. Cleveland, OH: The Lincoln Electric Company, 2000.

Figure Credits

Lincoln Electric Company, Module opener, 404F02-404F04

Salisbury Electrical Safety, 404F05

AWS D10.7M/D10.7:2008, Table 12, reproduced with permission of the American Welding Society (AWS), Miami, Florida, 404F10

Sumner Manufacturing Co., Inc., 404F13

All-Fab Corporation, 404SA01

Appendix A

Performance Accreditation Tasks

The Performance Accreditation Tasks (PATs) correspond to and support learning objectives in the *AWS EG2.0:2006 Curriculum Guide for the Training of Welding Personnel; Level I – Entry Welder.*

PATs provide specific acceptable criteria for performance and help to ensure a true competency-based welding program for students.

The following tasks are designed to evaluate your ability to run V-groove welds with GMAW equipment in three standard test positions using aluminum filler wire of the appropriate diameter and shielding gas. Perform each task when you are instructed to do so by your instructor. As you complete each task, show it to your instructor for evaluation. Do not proceed to the next task until told to do so by your instructor. For AWS 2G and 5G certifications, refer to *AWS EG3.0:1996 Guide for the Training and Qualification of Welding Personnel; Level II – Advanced Welder* for bend test requirements. For AWS 6G certifications, refer to *AWS EG4.0:1996 Guide for the Training and Qualification of Welding Personnel; Level III – Expert Welder* for bend test requirements.

V-GROOVE WELDS ON ALUMINUM PIPE WITH BACKING IN THE 2G POSITION

Using aluminum filler wire of the appropriate diameter, proper shielding gas, and stringer beads, make V-groove welds on aluminum pipe in the 2G position.

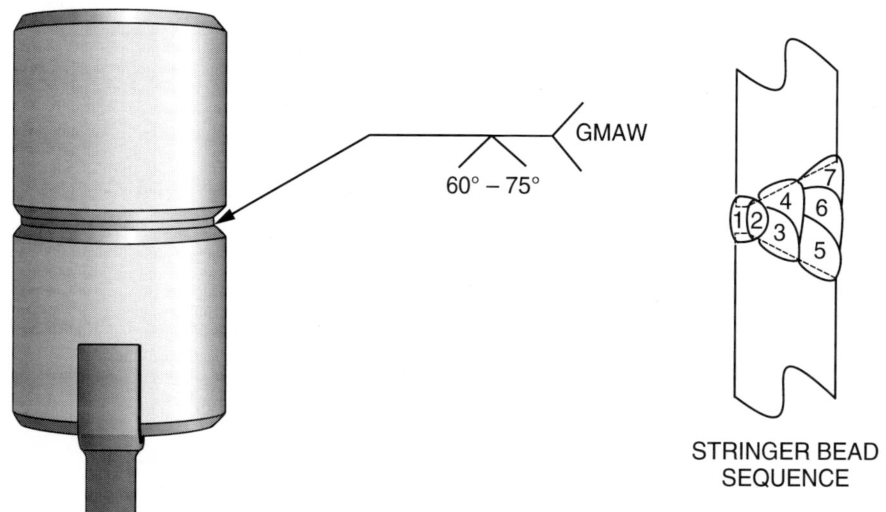

Criteria for Acceptance

- Uniform appearance on the bead face __________
- Craters and restarts filled to the full cross section of the weld __________
- Acceptable weld profile in accordance with the *ASME Boiler and Pressure Vessel Code – Section IX* __________
- Smooth transition with complete fusion at the toes of the weld __________
- Complete uniform root reinforcement at least flush with the inside of the pipe to a maximum of ⅛" __________
- No porosity __________
- No excessive undercut __________
- No cracks __________
- No overlap __________
- No incomplete fusion __________

V-GROOVE WELDS ON ALUMINUM PIPE WITH BACKING IN THE 5G POSITION

Using aluminum filler wire of the appropriate diameter, proper shielding gas, and stringer or weave beads, make V-groove welds on aluminum pipe in the 5G position.

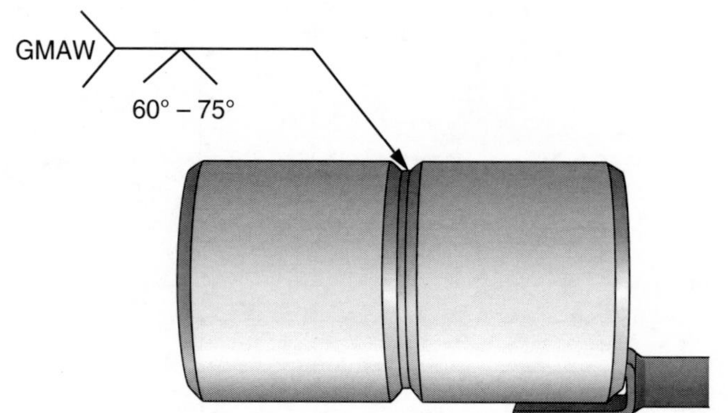

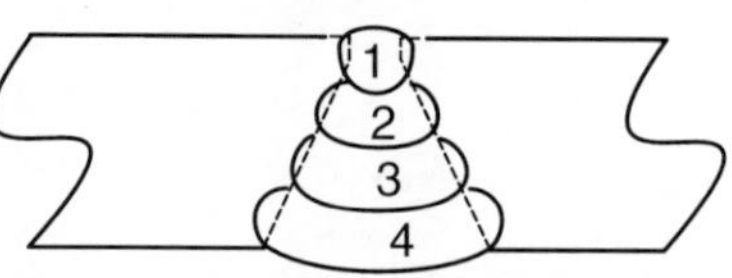

WEAVE BEAD SEQUENCE

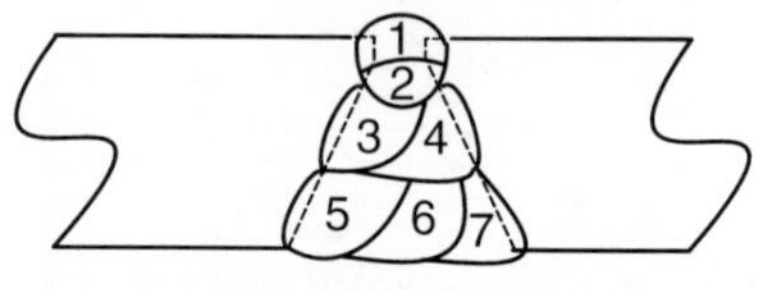

STRINGER BEAD SEQUENCE

404A02.EPS

Criteria for Acceptance

- Uniform appearance on the bead face ______
- Craters and restarts filled to the full cross section of the weld ______
- Acceptable weld profile in accordance with the *ASME Boiler and Pressure Vessel Code – Section IX* ______
- Smooth transition with complete fusion at the toes of the weld ______
- Complete uniform root reinforcement at least flush with the inside of the pipe to a maximum of ⅛" ______
- No porosity ______
- No excessive undercut ______
- No cracks ______
- No overlap ______
- No incomplete fusion ______

V-GROOVE ON ALUMINUM PIPE WITH BACKING IN THE 6G (OR 6GR) POSITION

Using aluminum filler wire of the appropriate diameter, proper shielding gas, and stringer beads, make V-groove welds on aluminum pipe in the 6G (or 6GR) position.

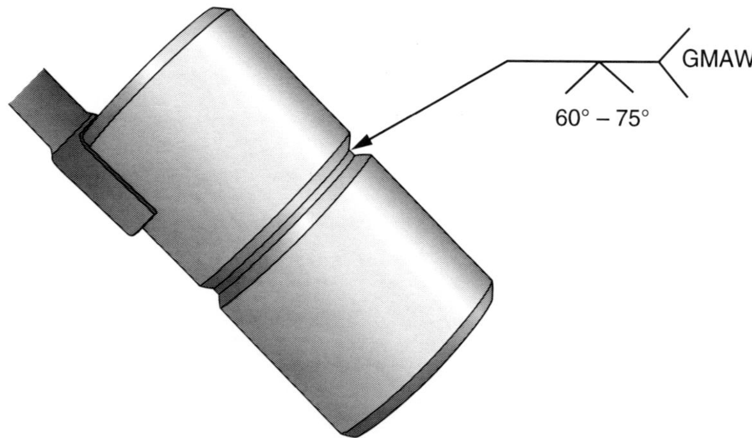

NOTE: IF REQUIRED FOR QUALIFICATION PURPOSES, A RESTRICTING RING MAY BE ADDED TO THE 6G POSITION COUPON TO FORM A 6GR POSITION COUPON.

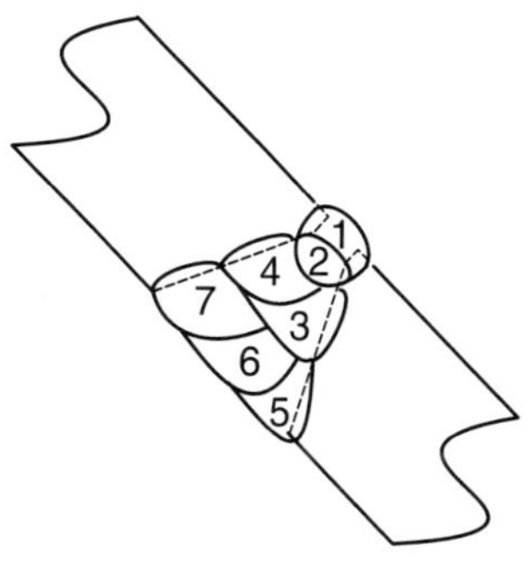

404A03.EPS

Criteria for Acceptance

- Uniform appearance on the bead face __________
- Craters and restarts filled to the full cross section of the weld __________
- Acceptable weld profile in accordance with the *ASME Boiler and Pressure Vessel Code – Section IX* __________
- Smooth transition with complete fusion at the toes of the weld __________
- Complete uniform root reinforcement at least flush with the inside of the pipe to a maximum of ⅛" __________
- No porosity __________
- No excessive undercut __________
- No cracks __________
- No overlap __________
- No incomplete fusion __________

NCCER CURRICULA — USER UPDATE

NCCER makes every effort to keep its textbooks up-to-date and free of technical errors. We appreciate your help in this process. If you find an error, a typographical mistake, or an inaccuracy in NCCER's curricula, please fill out this form (or a photocopy), or complete the online form at **www.nccer.org/olf**. Be sure to include the exact module ID number, page number, a detailed description, and your recommended correction. Your input will be brought to the attention of the Authoring Team. Thank you for your assistance.

Instructors – If you have an idea for improving this textbook, or have found that additional materials were necessary to teach this module effectively, please let us know so that we may present your suggestions to the Authoring Team.

NCCER Product Development and Revision

13614 Progress Blvd., Alachua, FL 32615

Email: curriculum@nccer.org
Online: www.nccer.org/olf

❑ Trainee Guide ❑ AIG ❑ Exam ❑ PowerPoints Other ____________________

Craft / Level: Copyright Date:

Module ID Number / Title:

Section Number(s):

Description:

Recommended Correction:

Your Name:

Address:

Email: Phone:

Glossary

Anodic coating: An artificial buildup of aluminum oxides on the surface of aluminum alloys that improves corrosion resistance. This coating is many times thicker than a coating of naturally occurring oxides.

Coalescence: The growing together or growth into one body of the materials being joined.

Helium/argon mixture: A shielding gas mixture in which helium is added to argon to raise the temperature of the arc, which promotes higher welding speeds and deeper weld penetration.

Ingot: A mass of metal shaped for convenient storage and transport.

Liquation: The process of separation by melting, such as in an alloy when one element melts while the others remain solid.

Index

D

E

F

G

Z